JURASSIC

208 million to 145 million years ago. Pangaea slowly split in two. The climate became drier with more seasons. The first bird appeared. So did giant plant-eating dinosaurs called sauropods. They walked on four legs. Many kinds of theropods, meat-eating dinosaurs, also developed. They walked on their two hind legs and had hollow bones.

CRETACEOUS

145 million to 65 million years ago. The first flowers appeared. The land divided into several continents with many different climates. Some dinosaurs lived in deserts. Others survived in hot swamps or Arctic cold. Most of the dinosaurs we know come from the Cretaceous. They include not only theropods and sauropods but duckbilled and horned dinosaurs. One of the last theropods was the biggest dinosaur hunter, *Tyrannosaurus rex*.

335 million years ago	230 million years ago	225 million years ago	145 million years ago	65 million years ago	1.5 million years ago

First reptile

First pterosaur

First dinosaur

First bird

End of dinosaurs

First man

SIMON & SCHUSTER

NEW YORK

LONDON

TORONTO

SYDNEY

TOKYO

SINGAPORE

Kings
of Creation

**HOW A NEW BREED OF SCIENTISTS IS
REVOLUTIONIZING OUR UNDERSTANDING
OF DINOSAURS**

by Don Lessem

Illustrated by John Sibbick

SIMON & SCHUSTER
Simon & Schuster Building
Rockefeller Center
1230 Avenue of the Americas
New York, New York 10020

Copyright © 1992 by Don Lessem

Designed by Liney Li
Manufactured in the United States of America

Library of Congress Cataloging-in-Publication Data
Lessem, Don.
Kings of creation : how a new breed of scientists is revolutionizing
our understanding of dinosaurs / by Don Lessem ;
illustrated by John Sibbick.
p.   cm.
Includes bibliographical references (p.   ) and index.
1. Dinosaurs. 2. Dinosaurs—Research. I. Title.
QE862.D5L49   1992
567.9'1—dc20   91-47904   CIP
ISBN 0-671-73491-1

ACKNOWLEDGMENTS

I am indebted to those dinosaur researchers who so freely gave of their time, their research, even the sandwiches out of their larders, and on one occasion, the shirts off their backs in the task of providing me some remedial education as well as giving the book its shape and much of its content. First thanks go to Bob Bender, friend and editor past and future, and to Peter Dodson and David Weishampel, for helping me define appropriate themes. And to Stephen Jay Gould and Robert Bakker for their helpful preliminary advice regarding this endeavor.

Once begun, I would not have gotten far were it not for the kindness and expertise of the following, in no particular order: Mark (pronounced "Mahk") Goodwin, Dale Russell, David Norman, Mark Norell, Mike Novacek, Stephen and Sylvia Czerkas, Connie Barut, Ted Daeschler, Tony Fiorillo, Jack Horner, Patrick Leiggi, Carrie Ancell, Ray Rogers, Bea and Betsy Taylor, Tom Rich, Martin Sander, Carole Gee, Diane Gabriel, Paul Olsen, Hans Sues, Ken Carpenter, Brooks Britt, Bill Amaral, Neil Shubin, Rolf Johnson, Jim Jensen, Jim Farlow, Kay Behrensmeyer, Paul Sereno, Kevin Padian, William Clemens, Robert Spicer, Leo Hickey, Kirk Johnson, Jack Wolfe, Scott Wing, Andrea Krumhardt,

Gail Nelms, Howard Hutchinson, Roland Gangloff, Sam Tarsitano, Sankar Chatterjee, Kevin Small, Beth and Greg Henry, Walter Bock, Michael Parrish, Jacques Gauthier, S. M. Kurzanov, Eric Buffetaut, Halszka Osmolska, Keith Rigby Jr., Robin Bates, Doug Jones, Peter Galton, Jack McIntosh, Brian Noble, and the staff of the Ex Terra Foundation, Dong Zhiming and his associates at the Institute of Vertebrate Paleontology and Paleoanthropology in Beijing, Xiao Xi Wu of the Chongqing Natural History Museum, Su Fenglin and the staff of the Academia Sinica, Daniel Grigorescu, Academicians Artangerel Perle and Rinchen Barsbold and officials of the Republic of Mongolia, Philip Currie, Clive Coy, Don Brinkman, Pat Lee and the faculty and staff of the Royal Tyrrell Museum, the Society of Vertebrate Paleontology, Parker and Catana Van Hecke, Robert Dunlop, Nick Hotton, Arnie Lewis, Ken Stadtman, Wade Miller, Clifford Miles, Andrew Leitch, Bill Law (of Lessem-Law Joint Ventures, Ltd.), Larry Martin, Colin Pennycuick, Blue Magruder and the staff of Earthwatch, David Gillette, Jim Madsen, Ann Hollister, Sally Erickson and the staff of the Burke Museum, Karl Hirsch, and the librarians of the *Boston Globe* and Harvard's Museum of Comparative Zoology.

I owe a second deep debt to Peter Dodson, who served as technical adviser and proofreader supreme for the entire text.

I owe my life to skilled pilots, among them Red in Prudhoe Bay, and the unnamed heroes who managed to keep the Mongolian commercial air fleet, such as it was, operational.

My return to my first scientific flight of fancy, the dinosaurs, would not have been possible without the professional journalistic encouragement I received to do so, first from my most frequent, insightful, and patient editors, Ande Zellman, Nils Bruzelius, and Gerry O'Neill at the *Boston Globe.* For years I have been indebted to these editors and their superiors, Jack Driscoll and Ben Taylor, for the extraordinary opportunities they have provided me as a free-lancer to hang my hat and hoist my byline at that paper. Particular concern and patience in that regard have been provided me by Don Skwar and his staff of the *Globe*'s nonpareil sports department. Editors at several magazines, particularly Sandra Shaul of *Rotunda,* Paul Hoffman of *Discover,* Joe Poindexter of *Life,* Kevin McKinney of *Omni,* and Kit Carlson of the *Discov-*

ery Channel Magazine, were kind in supporting my dinosaur journeys. Many thanks for their interest in dinosaurs and my work go to Paula Apsell, Mark Davis, Alain Jehlen, Bill Grant, Evan Haddingham, Queenie Coyne, and others at "Nova."

And I would not have had the luxury of investigating this topic far enough to visualize a book worth writing without the largess of the Knight Science Fellowship at MIT. The fellowship is a unique opportunity for midcareer reflection and study whose inception and management owe largely to one journalist, Victor McElheny.

Nor could I have pursued the subject with such abundant resources without the expert representation of Al Zuckerman of Writers House, whose long-standing kind interest in my career can never be repaid.

I'm grateful also for the skills of S&S's samurai editor Leonard Mayhew, who made quick, adept, and painless (to me) cuts and added much focus to an overlong text. I am extremely fortunate that John Sibbick, one of the world's premier dinosaur illustrators, took the time and a pay cut to grace these pages with original art he designed and thoroughly researched entirely on his own. Designer Deb Perugi provided maps, time lines, and graphics. Paul Sereno, Beth and Greg Henry, Stephen and Sylvia Czerkas, and wizard free-lancer Jeff Frishman lent their slides. Martha Mueller and Michael Shear provided me the perspectives of highly intelligent lay readers.

A few last personal thanks: I received support and encouragement early on, even as a tiresome five-year-old and self-announced dinosaur expert, from my parents, brother, and Aunt Sylvia. My early interest in journalism was furthered by two outstanding high school teachers, Mr. Greenwood (damned if he'd tell us his first name) and Eric Rothschild, and two journalistic icons who kindly wrote back, Russell Baker and Frank Deford. As for natural-science tutelage, I've been fortunate to take an education at the feet of two of the most original and inspiring minds going (and at times, clashing): Stephen Jay Gould and E. O. Wilson.

I saved the best for last—my wife and children. Rebecca and Erica have shared my interest in dinosaurs, proving themselves dedicated and talented fossil-diggers. My wife, Paula, was, as always, both means and motivation for keeping family life and work going happily.

FOR "AUNT SEAL,"

SYLVIA LOEWENFISH,

WHO SHARED HER

UNQUALIFIED LOVE OF LIFE

WITH ALL OF US FORTUNATE

TO HAVE KNOWN HER

CONTENTS

Introduction: The Truth About Dinosaurs 15

Glossary of Essential Terms 23

Chapter 1: Approaching the Dinosaur Frontier 27

Chapter 2: The Search for the First Dinosaur.............................53

Chapter 3: The Case of the Flying Dinosaur 79

Chapter 4: The Dinosaur Takeover 103

Chapter 5: The Dragons of Dashanpu 123

Chapter 6: Heresies, Hot-Bloods, and Holes 149

Chapter 7: Land of Giants 167

Chapter 8: Dinosaurs at the Poles......................................199

Chapter 9: The Gobi Dinosaur Chase....................................225

Chapter 10: Eggs, Babies, and New Dinosaurs 257

Chapter 11: Dinosaurs at the Brink—and Beyond? 287

Author's Note..307

Appendix A: Dinosaur Fossils and Footprints 311

Appendix B: Where to See and Dig Dinosaurs 317

Appendix C: Recommended Reading 321

Sources 325

Index...353

Kings
of Creation

The Truth About Dinosaurs

A revolution is going on in dinosaur science, a revolution only dimly understood outside the field. A new generation of paleontologists around the world is fast remaking our understanding of how dinosaurs lived and died. Every seven weeks a new kind of dinosaur is named. Nearly half of all the dinosaurs ever found have been identified in the past twenty years, and the pace of those discoveries is accelerating around the world. Some of these newfound animals stretch the physical limits of dinosaur growth and of our imaginations. Other new discoveries are changing our views of the development, social life, intelligence, and mobility of the host of remarkable creatures we call dinosaurs.

Dinosaurs abounded for more than 160 million years. For much of that time, they were the dominant land animals around the globe. By virtue of their size and the duration of their reign, they can fairly be hailed as the most successful land animals in the history of life on earth. Their fossilized remains have been found on every continent of the modern earth. We know them best from the high plains of the North American West, the Gobi, and Patagonia, but researchers have located dinosaurs in less popularly expected places: Laos, Antarctica, Switzerland, New Jersey.

Yet in nearly two centuries of searching for them we know only about three hundred dinosaur kinds (or genera) from the entirety of their lengthy era. At least three times that many forms of dinosaurs, according to recent scientific estimates, remain to be found. And in all the world's museums, there are only 2,100 individual dinosaur specimens of any sort. Each of these creatures was part of a larger assemblage of dinosaur species, and they were by no means all contemporaries. Each fossil-bearing rock formation contains an entirely different assortment of animals, with no known species lasting more than a few million years, a modest fraction of dinosaur time. So a very different cast of dinosaurs likely lived during the huge gaps—in time, geography, and habitat—that remain within the geological and fossil record of the dinosaur age.

Contemporary dinosaur paleontology is closing some of those gaps quickly. From China and Patagonia come new discoveries filling in gaping holes in the fossil record. New dinosaur finds include animals half the length of a football field—the largest creatures ever to walk the earth—and others smaller than a chicken. And there is the promise of much more from unprecedently long time-lines exposed in these fossil-laden rocks.

In Alaska and Australia, dinosaur-age habitats of seasonal darkness and cold, new dinosaur discoveries cast suspicion on the "Nuclear Winter" scenario of dinosaur extinction. In Mongolia, South America, India, and western North America new dinosaur finds challenge accepted models of how continents formed.

At the same time, these and other discoveries in Transylvania and Montana support fundamental evolutionary concepts of how numbers and kinds of animal species change when habitats are altered.

From China, India, North and South America, researchers have found abundant evidence of dinosaur social life, detailing much about behaviors only dimly glimpsed before from Asian egg finds to North American footprint assemblages. Multiple trackways of dinosaur footprints newly found in Colorado, Canada, China, Brazil, and Australia give fresh indication that many duckbilled and horned dinosaurs were herding, migrating animals, and that the predators who pursued them were capable of speeds of up to twenty-five miles an hour. Abundant new discoveries of horned and duckbilled dinosaurs in arctic Alaska suggest these dinosaurs were either winter-hardy or migrants of enormous stamina. In the desert sands of northern China, Canadian researchers on the most ambitious dinosaur hunt ever undertaken discovered the huddled bodies of young armored dinosaurs, the first evidence that these dinosaurs were not solitary animals. In western Canada, Chinese researchers on the same international expedition located the braincase of the smartest dinosaur yet known, a six-foot-long predator as smart as an ostrich and far smarter than any other creature of dinosaur days, our mammalian ancestors included.

A new discovery of that smart dinosaur's close relation in China testifies to the linkage of Asia and North America at the end of the dinosaur era. At other ends of the earth new dinosaur discoveries provide evidence of changing alignments of continents in the age of dinosaurs.

From coastal Australia have come tiny but clever plant-eating dinosaurs with huge owlish eye sockets, animals well adapted for a hard and dark winter life in a then-antarctic landmass.

In Argentina, a host of odd dinosaurs have lately been discovered, from huge, bulldog-faced predators to titanic plant-eaters, some of them armor-sided. Other huge browsers, as yet unnamed, were equipped with a bizarre double row of tall spines down their backs. These oddities are products of South America's long isolation from the rest of the world. Conversely, the recent find of a familiar armored dinosaur in an entirely unfamiliar locale for dinosaur fossils, Antartica, reflects the reunion of southern lands with North America at the close of the dinosaur era. While new discoveries in the field have spurred the dinosaur renaissance, many recent advances have originated in laboratories and museum collections, from applying new techniques to existing fossils. CAT-scanning of dinosaur skulls too delicate to dissect has revealed peculiarly birdlike features in some predatory dinosaurs. X rays have supported claims that a plastered-over and long-neglected skull of what was thought to be a young, large predator is instead a previously unknown adult form of a "pygmy" tyrannosaur, a tenth the weight of *Tyrannosaurus rex*. Remote-sensing apparatus has helped detect dinosaurs in the field. And amino-acid mapping of some peculiarly well-preserved arctic dinosaur fossils offers the real, if remote, promise that dinosaur DNA may yet be remade in the laboratory.

These finds, detailed herein, are all quite new to science. The instigators of this revolution in our knowledge of dinosaurs are shockingly few in number—scarcely thirty scientists are digging dinosaurs anywhere in the world. They are miserably paid and, with a few celebrated exceptions, little recognized. While dinosaurs are big business, the worldwide budget for dinosaur exploration is less than $1 million a year, and only two American museums (The Museum of the Rockies and Yale's Peabody Museum) have a dinosaur paleontologist on staff. Like all those who work more for love than money, the new dinosaur hunters are engagingly unpretentious about their chosen field. Their number no longer includes the hard-living, unschooled field men and the autocratic academics of bygone days.

Today's dinosaur scientists are indeed young, for the most part, but they are not so much brash iconoclasts as multitalented researchers. A few are autodidacts, such as John R. "Jack" Horner of the Museum of the Rockies, who identified the first dinosaur nests in North America, Wesleyan University physicist John "Jack" McIntosh, whose avocation has made him the world expert on the giant sauropod dinosaurs, and Karl Hirsch, a German-born septuagenarian and ex–maintenance man who is now a leading authority on fossil eggshell.

All of today's dinosaur investigators, according to Harvard University paleontologist Stephen Jay Gould, "are far better educated and trained than their predecessors to evaluate the behavioral, environmental, and evolutionary significance of the fossils they discover."

Only the vaguest sense of this vibrant, if imperiled, international science and what it reveals of the dinosaur world has reached the dinosaur-adoring public. For all the attention we lavish on the most spectacular and successful of all earth animals, we seldom make the effort to see them as scientists do. Until lately, dinosaurs were commonly viewed, by public and paleontologists alike, as sluggish, swamp-bound, cold-blooded behemoths, the last of a moribund reptilian line.

Many of us are still caught in time-worn misconceptions of dinosaurs. We have loved dinosaurs ever since we discovered them. That fascination has endured, for dinosaurs are a unique blend of fantasy and reality, fantastic creatures that actually existed and are now safely dead.

We love dinosaurs as much as ever today, but we still don't know them. Most of us, according to a recent poll, believe dinosaurs lived at the same time as primitive humans. The few dinosaurs we do know have been lumped together in the public mind, melding drastically different times and habitats. "Brontosaurus" (properly called *Apatosaurus*) cavorts with duckbills and stegosaurs, while *Tyrannosaurus rex* gives chase. In truth, *Tyrannosaurus rex* lived closer in time to today's animals than it did to *Stegosaurus* and *Apatosaurus*. (There are metaphors aplenty to bring home the staggering length of life's history compared to our own, including Twain's analogy of human history to the paint on the top of the Empire State Building. My own comparison of dinosaurs' hegemony to our own is made on the length of an arm. Consider that all of life on land began somewhere around your shoulder. Dinosaurs evolved somewhere above your elbow and died out near your wrist. All of human evolution, from beginning to present, is to be found on the tip of your fingernail.)

Now, thanks to a few celebrated recent discoveries—of nesting, care-giving dinosaurs and a catastrophic extinction event—the public view of dinosaurs is shifting radically.

Dinosaurs are now presented in the media as lithe, hot-blooded progenitors of birds, killed by a blast from space. Their discoverers are made out to be young mavericks, tilting at the narrow-minded scientific orthodoxy.

This new view of the dinosaur world is as wrongheaded and over-simplified as the old, but we have little use for the more complicated realities of dinosaur life and science. And we have little patience with the slow pace of careful scientific research, the biases, disagreements, and many unresolved questions that bedevil dinosaur research as any scientific inquiry.

Whenever dinosaurs are dug up, they prove considerably more

complex than our past or present popular conceptions. For example, only the meat-eating dinosaurs have much in common with birds, and a controversial new find suggests their resemblance may be due instead to a common ancestor.

Indeed, most dinosaurs were smaller than automobiles, and many of them may have been as swift as the fastest humans. Many of these smaller dinosaurs could well have been hot-blooded, and as smart as any creatures of their day. Some migrated, some communicated noisily, some lived in mobile herds.

But other dinosaurs appear to have been solitary, lumbering, dim-witted, and quite capable of warming themselves in other ways than we warm bloods do. Some dinosaurs were caring parents of helpless young, but their contemporaries bore young that were independent from birth.

Some dinosaurs did live in fetid swamps. But others lived in near-desert environs, in evergreen forests, in cold arctic darkness. And while the impact of an extraterrestrial object may well have killed off the last of the dinosaurs, other evidence suggests that dinosaurs died out, from other causes as yet unknown, well before any such catastrophe.

But the ongoing revolution in dinosaur studies is not about naming new corpses, nor even finding new murder weapons. Rather it is a pioneering effort to understand the behavior and evolution of communities of dinosaurs and other organisms in varied environments, and to chart the dinosaurs' role in a long and eventful stretch of earth history. Dinosaur paleontologists have just begun writing on a still largely blank slate of dinosaur daily life—how, and in what relationship to other dinosaurs and contemporary organisms, did dinosaurs grow, live, diversify, prosper.

My aim in this survey is to impart the same sense I've received from two years on the road with dinosaur diggers from Mongolia to Alaska to Montana and Nova Scotia. Globe-trotting notwithstanding, my emphasis remains on North American dinosaurs and dinosaur workers, a bias that can be defended by the fact that a plurality of both are found on this continent. The United States may be slipping as a world leader, but it is still number one in dinosaurs.

Nonetheless, with the help of several generous paleontologist advisers, I've selected a representative worldwide sampling of current and significant dinosaur detective work, which I describe in the chapters that follow. There is no activity better for capturing the excitement of the scientific treasure hunt than prospecting with a paleontologist for clues to life's history. "The game," as Sherlock Holmes put it, "is afoot."

Glossary of Essential Terms

Any discussion of dinosaurs and their anatomy can rapidly become rough going for those unfamiliar with specific anatomical and taxonomic terms involved, many of them so tongue-tying they become ripe subjects for lampooning. (As done hilariously by Rick Meyerowitz and Henry Beard in their book, *Dodosaurs: The Dinosaurs That Didn't Make It*. They define not just the familiar *Thesaurus,* but *Doublediplodocus, Etceteratops, Preposterous,* and *Triceratush* among the Hysterischian or "trick-hipped" order of "ineptiles" that thrived during the Preposterous Era.) But the following terms are essential to an understanding of dinosaur science.

TRIASSIC—From 245 to 208 million years ago, the first of three periods of the Mesozoic era and the time when dinosaurs first appeared. The land masses of the earth were united in a supercontinent, Pangaea.

JURASSIC—From 208 to 145 million years ago, the second period of the Mesozoic. By its close, the heyday of the largest dinosaurs, a narrow seaway developed, separating Pangaea into a northern continent, Laurasia, and a southern continent, Gondwana.

CRETACEOUS—From 145 to 65 million years ago, the last period of the Mesozoic, during which flowering plants and duckbilled dinosaurs developed. The end of the Cretaceous is thought to have marked the extinction of the dinosaurs, by which time the continents had moved into positions much like those they occupy today.

SPECIES—A group of organisms that are similar in appearance and can interbreed.

GENUS (plural, *genera*)—A grouping of species with common characteristics.

FAMILY—A category ranking of like organisms between a genus and an order.

ARCHOSAURS—A major group of reptiles that includes dinosaurs, pterosaurs, and their common ancestors, living crocodiles, and according to some paleontologists, birds. Dinosaurs were distinct from many other archosaurs in several anatomical features, including ankle anatomy.

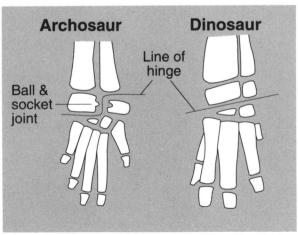

ORNITHISCHIAN—One of the two orders of dinosaurs, by traditional analysis based on hip anatomy. These are "bird-hipped" dinosaurs, which, like birds, have a pubis bone that lies parallel to another hip bone, the ischium. Stegosaurs and armored, dome-headed, and duck-billed dinosaurs are ornithischian.

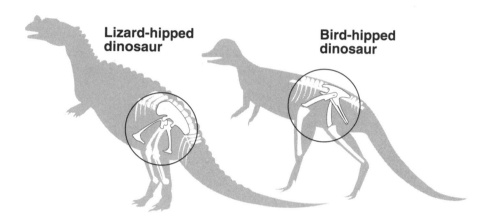

Lizard-hipped dinosaur

Bird-hipped dinosaur

SAURISCHIAN—The second order of dinosaurs, by traditional analysis based on hip anatomy. These "lizard-hipped" dinosaurs had a long pubis bone pointing forward and down from the hip socket. The meat-eating theropods and the giant herbivore sauropods are saurischian.

SAUROPODS—Large plant-eating saurischian dinosaurs, including the largest dinosaurs known, more than one hundred feet long.

THEROPODS—Largely bipedal, predatory ornithischian dinosaurs from smaller than man-sized hunters to forty-foot *Tyrannosaurus rex.*

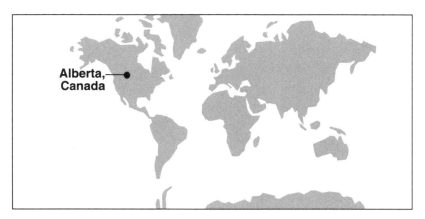

The World Today

CHAPTER 1

Approaching the Dinosaur Frontier

Two hours southeast of Edmonton the road dips out of the vast Alberta grainfields into a maze of narrow canyons. Walls of bare, banded gritty rock—dark red, black, and chalk white in a layer cake of mudstone, sandstone, and coal—zigzag through the valley. Erosion has sculpted strange features here, fashioning rock pillars to adorn the moonscape. Some, vaguely human in outline, were thought by the native Americans to be petrified giants. They are still known to the locals as hoodoos (from voodoo). The misshapen titans recall the banks of a surging river of glacial meltwater that formed them, this canyon, and the myriad cliffs, just thirteen thousand years ago. That was the close of the last ice age, a mere blink in geologic time. The terrain that this postglacial erosion has exposed, however, is older than the Rocky Mountains. It is as old as the dinosaurs. And embedded in the ancient rock are remains of the subtropical river-delta environment in which dinosaurs lived some 69 million years ago, preserved as nowhere else on earth. This is the world's greatest dinosaur hunting ground.

From nearly a century of prospecting in these badlands, two hundred miles north of the Montana border, have come many spectacularly complete skeletons of duckbilled dinosaurs. They are among twenty species of dinosaurs unearthed near the tiny city of Drumheller, nestled in the badland ravines on the banks of the Red Deer River. This marvelous menagerie includes tyrannosaurs, dome-headed dinosaurs, armored dinosaurs, and duckbills adorned with various elaborate crests.

And within two hours' drive, in the badlands of Dinosaur Provincial Park, more dinosaur species are known—thirty-seven have been identified—than in any other locality on earth. What Dinosaur Provincial Park has yielded is staggering: more than three hundred articulated (connected) skeletons of dinosaurs and more than ninety newly identified kinds of fish, amphibians, reptiles, and mammals.

If you are looking for dinosaurs, and those people who study them, badlands are the best place to start. The bands of rock visible in their hillsides mark layers of time, eras when meandering streams brought minerals into ancient deltas and valleys. Over thousands, even millions, of years, the accumulating sediments compressed into layers of rock. Where that silt and sand settled relatively quickly over bones and carcasses, eggs, plants, and footprints might not have eroded. Minerals might enter them, creating fossils. The heaving and shifting of the earth's crust pushes fossil-bearing rocks near the surface. If you locate strata of rock from land of dinosaurian age—225 million to 65 million years old—the prospects are good you'll find dinosaurs in them.

But only in badlands, where an environment of deposition has now shifted to one of erosion, are dinosaurs readily found. Elsewhere, if dinosaur-aged rock is near the surface, it is obscured by vegetation and the trappings of civilization. Occasionally, when a mine or a quarry is opened in developed and well-vegetated lands from Switzerland to Szechuan to New Jersey, dinosaurs are exposed. In the badlands, winds sweep over bare rock, eroding the surface and exposing, here and there, the remains of dinosaurs. Every day, in such barren places, dinosaurs pop up. If they are not discovered, they, too, will erode and in a matter of decades turn to dust.

So it is to these stark and extreme realms that dinosaur paleontologists have ventured for more than one hundred years. Often they follow in the footsteps of amateur fossil finders fortunate to happen upon, and wise enough to report, a discovery. And often they make their own discoveries, ambling up and down rocky hillsides, or crawling along ancient stream channels on their hands and knees until they come upon a telltale piece of bone.

As it has always been, dinosaur-hunting is an amalgam of art, science, and serendipity. "We're like coroners," Smithsonian paleontologist Michael Britt-Surman told *Newsweek* magazine, "except with us the crime occurred more than 65 million years ago, all the witnesses are dead and all the evidence has been left out in the rain."

The strange saga of one of the most recent and significant of discoveries in the Drumheller badlands reveals at once how sophisticated dinosaurs were, and how sophisticated dinosaur science has become, while still resting on an unsteady foundation of simple luck.

The Smartest Dinosaur

Many dinosaurs are so poorly known, they have been identified by only a few shed teeth, a tradition fraught with scientific problems. From just a few pointy teeth, *Troödon formosus* (literally "wounding tooth," the first word pronounced *True-o-don*) was one of the first dinosaurs named, more than a century ago. Those teeth indicated that *Troödon* was some sort of small hunter from the latter days of the dinosaurs, 75 million years ago. But for a century not much more of *Troödon* than its teeth were known.

That cloud of uncertainty began to clear one spring afternoon in 1983 in the badlands on the outskirts of Drumheller. Jack Horner was up from Montana State University, visiting his friend and fellow dinosaur paleontologist Phil Currie. Currie was looking over the progress of construction of his sponsoring institution, the Royal Tyrrell Museum of

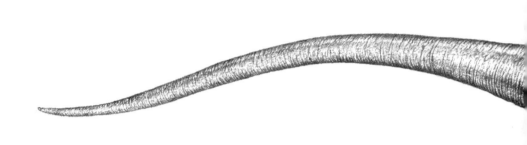

© SIBBICK

Troödon *[John Sibbick]*

North America's leading dinosaur paleontologists gather annually at the Society of Vertebrate Paleontology annual meetings. Several are pictured here in 1988 at the SVP conference at the Royal Tyrrell Museum in Alberta, Canada. Between a Tyrannosaurus rex *and a horned dinosaur skull are, left to right rear, Robert Bakker, John McIntosh, Paul Sereno, David Weishampel, Philip Currie, Peter Dodson; left to right foreground, John R. Horner, Sankar Chatterjee, David Gillette. [Photo courtesy Rich Frishman]*

Paleontology—then nearing completion as the world's largest fossil museum. The two scientists strolled through the badlands behind the construction site, heads down in characteristic paleontologist's posture.

Currie is an outstanding fossil finder. But Horner is even better. He is widely regarded by his peers as one of the best "field men" in the world. On this day Horner spotted a fossil and picked it up, quickly recognized it as the jaw of a small carnivorous dinosaur, and showed it to Currie. As the world expert on dinosaur teeth, Currie has examined thousands of dinosaur teeth worldwide. Just as Sherlock Holmes could identify more than three hundred forms of cigarettes by their ash alone, Currie is uniquely able to identify a dinosaur from a single tooth. Those in the jaw Horner picked up, Currie recognized as teeth attributed to *Troödon* and to another small predator. The rest of the jaw, three or four inches long, appeared to be jutting from the hillside.

Currie planned to come back the next day with a crew and excavate the find. Only, as Currie wincingly recalls, "it rained. For a week. When we were able to get back there, it was gone. We couldn't find it."

And so the *Troödon* puzzle might have remained unsolved had not Horner returned when the museum opened in 1985 for another visit, and another walk with Currie. Again he found the same jaw, a feat as unlikely as locating a needle in a haystack.

This time, Currie and Horner flagged the site with an orange marker. When Currie and his crew had excavated, prepared, and studied the

Hidden in the badlands outside Drumheller, Alberta, the Royal Tyrrell Museum of Paleontology is the world's largest fossil museum and a center of dinosaur studies. [Photo courtesy Rich Frishman]

jaw, Currie found no fewer than four basic tooth types, teeth formerly thought to come from four different kinds of dinosaurs.

The following year, 1986, while prospecting with Currie in Dinosaur Provincial Park, a Chinese fossil preparator, Tang Zhilu, found the bony braincase of a *Troödon*. The size of an avocado, it was remarkably large for a dinosaur no bigger than a man, offering compelling evidence of its intelligence. This case held a brain nearly three times larger than that of any other dinosaur then known, and larger than any creature of its day, mammals included.

More kinds of dinosaurs have been found here in the badlands of Dinosaur Provincial Park in Alberta than anywhere else on earth. [Photo by author]

Statistical measurements have correlated brain size with intelligence in living animals. To judge from its brain size and by estimated brain-to-body-weight ratio, *Troödon* was at least as smart as an ostrich. And by these standards, *Troödon* appears to have been bright enough to coordinate not only its own keen stereoscopic vision and gripping hands, but to hunt cooperatively as many advanced predators do today. Little troödontids, not huge tyrannosaurs, may have been the most efficient hunters of dinosaur times—fast, agile, and above all, clever.

In the few years since Horner and Currie made their discovery and

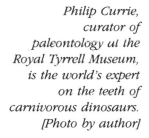

Philip Currie, curator of paleontology at the Royal Tyrrell Museum, is the world's expert on the teeth of carnivorous dinosaurs. [Photo by author]

rediscovery, the life history of *Troödon* has become far clearer. Horner found *Troödon* eggs and skeletal remains in Montana, establishing *Troödon* as less than six feet in length and one hundred pounds in weight. Currie and Canadian and Chinese colleagues found close cousins of *Troödon* in Mongolia, confirming that the animal had spread across the then-linked continents of Asia and North America. And they found bones of a member of the *Troödon* family 30 million years more ancient, making the *Troödon* family one of the most long-lived known from dinosaur times.

Currie invalidated many long-standing dinosaur names by detailing that the Cuisinart selection of tooth blades belonged to a single animal, *Troödon*. And by detailed studies of the braincase, including CAT scans, Currie established that *Troödon* was the most birdlike of all dinosaurs in its cranial structure. If dinosaurs were the ancestors of birds (see Chapter 3), the long-mysterious *Troödon* now provides one of the best pieces of supporting evidence.

Horner, Currie, and *Troödon* all play central roles in the dinosaur science revolution. But to grasp how far that quest has advanced in recent years, one must first appreciate its richly colored, if unevenly productive, past. It is a history worthy of long accounts, and several have been written (see Appendix C for recommendations).

Discovering Dinosaurs

In brief, people around the world were digging up dinosaurs for many centuries before scientists created a term for the finds. To their unknowing discoverers, these fossils were dragon bones, remains of giants, and biblical creatures. The first scientific description of a dinosaur bone, albeit unwitting, was of a massive thighbone identified as the scrotum of a giant by Oxford University chemistry professor Robert Plot in 1677.

Among the first to recognize vertebrate fossils as the petrified remains of long-extinct actual creatures was Baron Georges Cuvier. As a young anatomist, in 1796, Cuvier examined the jaws of a large fossil animal excavated outside Paris. It was not a dinosaur, but a *Paleotherium*, a close relative of modern horses. Cuvier's contemporaries, Thomas Jefferson among them, had argued extinction could not, in principle, occur. But with *Paleotherium* and other fossil finds, Cuvier established, as Stephen Jay Gould puts it, "a lengthy, continuous, and sensible history of the earth." His was a theoretical model that could accommodate dinosaurs.

Bones of dinosaurs were found in the early 1800s in Sussex, England, by several individuals, supposedly including Mary Ann Mantell, wife of physician and amateur fossil collector Dr. Gideon Mantell.

As legend has it, Mrs. Mantell saw her first dinosaur fossil, a plant-eater's grinding tooth, in a pile of stones used for road repair. More likely the dinosaur tooth turned up at the Mantells' front door, in a shipment of fossils Mantell bought from quarrymen in Tilgate Forest. Dr. Mantell described the dinosaur's tooth in 1822 and named the animal from which it came *Iguanodon* in 1825, identifying it incorrectly as an extinct relative of the iguana. In 1824, William Buckland of Oxford University, famed anatomist and geophagist (he enjoyed consuming exotic animals), named the predator *Megalosaurus* from a partial jawbone discovered in the north of Oxfordshire.

The first to use the term *dinosaur* ("terrible lizard") was British anatomist Sir Richard Owen, who named the Dinosauria in 1841 to encompass these giant finds. He saw them all as lumbering, elephantine beasts—a powerfully enduring misconception.

That image was altered, at least in one respect, by the first American dinosaur paleontologist, Joseph Leidy of Philadelphia. In 1858, he named the duckbilled *Hadrosaurus* from a skeleton found in a New Jersey marl pit. This dinosaur had hind limbs far longer than its front legs, suggesting a stance more like that of a kangaroo than an elephant.

A public and scientific mania for dinosaurs began in the United States in the late 1870s. The Great Dinosaur Rush, as it is often called, was fueled by two wealthy, stubborn, and self-aggrandizing Easterners, Edwin Drinker Cope of Philadelphia and Othniel Charles Marsh of Yale. In 1877, each was independently sent dinosaur bones from a rich dinosaur quarry in Colorado, found by two schoolmasters. The mailing proved an unfortunate coincidence, as Marsh and Cope were already enemies. Marsh had embarrassed Cope by pointing out that Cope had mounted the head of a plesiosaur on the wrong end of the animal.

As rival collectors, Cope and Marsh paid explorers to plumb the fossil treasures of the American West. As a result, both men were presented with hundreds of giant bones. The rivalry was heated. Prospectors fought off Indians, fired at each other, hid their finds, and appropriated their rivals' discoveries. Their employers battled in the press, accusing each other of plagiarism and libel. Nearly every find was dubbed a new species. Many of the 136 dinosaur species Cope and Marsh named over the next two decades have since been invalidated. Yet some are among the best-known dinosaurs today, for the Cope and Marsh dinosaurs include *Stegosaurus, Diplodocus,* and *Apatosaurus* (formerly "Brontosaurus").

At the same time as Cope and Marsh were entering the dinosaur game, a phenomenal dinosaur find was made in Europe. Miners probing a coal seam in Bernissart, Belgium, in 1878 came upon thirty-nine skeletons of *Iguanodon,* many of them complete.

During the first half of the twentieth century, dinosaur discoveries proliferated throughout the world. From Argentina to East Africa, across Europe, Asia's Gobi Desert, China, India, and Australia, European and American paleontologists uncovered dramatic and unfamiliar dinosaur fossils.

The pace of dinosaur exploration slowed, most noticeably in the United States, beginning in the Depression years of the 1930s. In the postwar years few scientists entered the field, and little in the way of fresh research emerged.

For a century, until the late 1960s, the business of dinosaur paleontology was the collecting of new dinosaurs, preferably of museum display quality. Theories of dinosaur life and death changed little, though curiously, some early dinosaur paleontologists saw the animals as active, agile creatures.

To most scientists, dinosaurs were so ancient that they had to be primitive. Since their fossils were found in stream channels, the dinosaurs must, the scientists reasoned, have lived by and in the water. Water would give them buoyancy to support their enormous bulk. Their brains were small, compared to ours, so they must have been stupid. This stupidity, or some other failing, caused them to become extinct. Our ancestors, the mammals, triumphed.

Even the field techniques changed little over the century. To this day, newfound fossils are still coated with shellac (now polyvinyl acetate), wrapped in tissue paper, and swaddled in plaster-soaked burlap just as they were by Marsh's field men a century ago. Paleontologists still lick puzzling fragments to determine if they are fossils. (With its porous structure bone sticks to the tongue. Rock does not.)

What has changed, in today's renaissance of dinosaur studies, is what is made of those bones. And if one locale can be singled out as a nexus for the remaking of dinosaur studies, it is the badlands of southern Alberta in western Canada.

Discovering a Dinosaur World

In 1884, on the banks of the Red Deer River in what is now Drumheller, geologist Joseph Burr Tyrrell happened upon the imposing skull of a carnivorous dinosaur, *Albertosaurus*. Soon, the Canadian Dinosaur Rush was on.

The region produced a bonanza of fossils for early-twentieth-century collectors who scoured the banks of the Red Deer River and rafted their treasures downriver to Drumheller. Among the collectors were the Sternberg family, ace fossil finders, and the legendary Barnum Brown of the American Museum of Natural History, the man who discovered *Tyrannosaurus rex*. The rival explorers found duckbilled, horned, and predatory dinosaurs, previously unknown and marvelously preserved.

Yet, after the Canadian Dinosaur Rush ended in 1917, paleontologists all but ignored Drumheller and nearby Dinosaur Provincial Park, until American-born Canadian paleontologist Dale Russell and his student, Canadian-raised American paleontologist Peter Dodson, began exploring the park in the 1960s.

Dodson is a gentle, soft-proportioned man with sad eyes and an auburn goatee. His high, lilting voice brings to mind a pediatrician or a pastor. Indeed, he considered the first as a career and is a devout Catholic. But his precise diction and attention to detail are dead giveaways of a career academic. Dodson is a teacher of anatomy at the University of Pennsylvania. Within the dinosaur fraternity he is esteemed as a well-rounded scholar and an excellent field researcher, respected for his environmental studies of Montana dinosaurs, his theories on gender-related differences in duckbill dinosaur anatomy, and for his first piece of field research, at Dinosaur Provincial Park.

Dodson was inspired to use the dinosaur fauna there for a pioneering study of the circumstances of dinosaur death and preservation—taphonomy. It was the first time this then little-known methodology, developed by a Soviet paleontologist more than a decade before, had been applied to dinosaurs. Dodson sought to examine fossils to understand not just which dinosaurs were which, but how they all fit into their environment, in life and in death.

Seventy-five million years ago, Dinosaur Provincial Park was more like a southern bayou than the maze of canyons cut in a northern plain it has lately become. Pelting rain and melting snow on the newly upthrust Rocky Mountains of 76 million years ago washed thick layers of sand and mud eastward onto the deltas of a shrinking inland sea, a vast seaway that had once covered most of interior North America from the Arctic to the Caribbean. By geological time standards, those layers of sediment were quickly, if not thickly, deposited. Yet they did continue building what is now the rock of the Judith River Formation for millions of years, attaining a thickness of more than eight hundred feet.

In today's Dinosaur Provincial Park those rocks are exposed, sculpted into seven thousand badland acres. These waves of orange, rust, and ash-gray striated hoodoos and cliffs form so spectacular a primeval landscape that the makers of a caveman film, *Quest for Fire,* did their shooting here. The Red Deer River winds through the center of the park, bordered by a thin row of aspens and cottonwood frequented by more than a hundred species of birds from golden eagles to prairie falcons, kestrels and rock wrens, and herds of whitetail and mule deer. Beyond the river margin, the land is bare, dusty, and forbidding.

Seventy-five million years ago, this was a far more hospitable place. Transported soil swept onto a broad flatland, a marshy stretch dotted with lakes and slow-moving streams. Turtles and crocodiles hid among the rushes and cattails, tiny mammals scampered beneath the dawn redwood and sycamore saplings in a dinosaur-dominated landscape that looked much like the familiar museum dioramas of prehistoric habitats, complete with ferns, palms, and other flowering trees. Helmeted duckbills foraged along the sandbars, as did squat armored dinosaurs and rhino-sized horned dinosaurs, all likely chased by the huge and frightening *Albertosaurus.*

An Albertosaurus *stands over its prey. The twenty-five-foot-long carnivore was first discovered in Drumheller. This skeleton is one of the many mounted casts of dinosaurs at the Royal Tyrrell Museum. [Photo courtesy Royal Tyrrell Museum]*

Dodson discerned much of this by analyzing not only fossils, but stream channels that held the bones, even annual growth rings in crocodile fossils and petrified wood that testified to the temperate seasonality of the dinosaur-age climate.

For his analysis of how the dinosaurs themselves died and were buried, Dodson catalogued the positions of the thousands of dinosaur bones he and others found. As detectives chalk the outline of a corpse in recreating the circumstances of a murder, Dodson mapped the dismembered bodies of some twenty-five varieties of dinosaurs. He located them only rarely in the overbanks but often in the channel sediments. The bigger dinosaurs were buried in sand along an east-west axis, parallel to the flow of ancient water. Bodies that reached their final resting place in the bows or side channels were often set perpendicular to the current, perhaps blown there by the wind, and were better preserved. Dodson concluded that the dinosaur fossils in the park had been transported by water. How different bodies of water sorted and acted upon dinosaur bones was a finding with universal applications.

Some of Dodson's other conclusions were, by his own later omission, probably erroneous—he reasoned that the most abundant dinosaurs, the duckbills, may have lived in the water, as neither entirely aquatic nor terrestrial animals. Dinosaur workers, Dodson included, now see duckbills as herding, land-loving herbivores (though footprints indicate some could swim).

The most significant application of taphonomy into dinosaur studies would not come for another decade, in this same park. Currie, not Dodson, would be the one to do it. But in his pioneering attempt to bring new scientific techniques to bear on understanding dinosaurs' environments as well as their individual anatomies, Dodson had brought dinosaur science into a new age.

The modern sense of how dinosaurs lived and behaved can be traced to another chance discovery made in the 1960s, across the Albertan border in the familiar dinosaur-hunting badlands of Montana.

Dinosaurs Dance and Die

Prospecting in central Montana in 1964, Yale paleontologist John Ostrom happened upon the long limbs and prodigious claws of a man-sized predator he named *Deinonychus* ("terrible claw"). He concluded, as did others who examined the fossils and read his papers, that this and at least some other dinosaurs were highly active, perhaps even warm-blooded, and descended from the dinosaurian ancestors of all birds. With the sweep of a killing claw, the scientific and public perception of dinosaurs was radically altered.

Warm-bloodedness became a cause célèbre in the 1970s in the talented hands of Ostrom's brilliant and controversial student, Robert Bakker. Bakker's spirited advocacy of fast-moving, warm-blooded dinosaurs as a separate class of organisms distinct from reptiles prompted a surge of research and speculation on dinosaur behavior and physiology. The inquiries were spurred on by Bakker's allegations that dinosaur paleontology was a hopelessly hidebound orthodoxy.

But the lines of evidence forwarded by Bakker and pursued by others often proved blind alleys. Bakker saw canal-like structures within dinosaur bones as common to fast-growing, warm-blooded creatures. Others showed the canals to be features of large living animals, both warm- and cold-blooded. Bakker estimated the ratio of predators to their prey in dinosaur times and found proportions consistent with those of warm-blooded, mammalian killers and victims in the African savanna today. Others points out that the fossil evidence for dinosaurs was so biased, incomplete, and inadequately computed that Bakker's ratios were suspect.

Ongoing studies of bone structure and growth rates based on a wealth of new fossil evidence may yet reopen the debate over dinosaur hot-bloodedness (see chapter 6). But for most researchers the case of the hot-blooded dinosaurs was closed with a symposium in 1978. The verdict: smaller and more active dinosaurs were likely "warm-blooded," though the term hardly begins to define the various heating and cooling strategies pursued by animals past and present. Many dinosaurs may have had some metabolic features like our own, at least in their youth. As for larger dinosaurs, their own body mass could effectively warm them without requiring energy from the consumption of huge quantities of food. For all dinosaurs, metabolism, paleontologists agreed, is a function we may never be able to puzzle out from fossils.

Though it reached a dead end, the hot-blooded dinosaur debate captivated public interest anew in dinosaurs, revamped the animals' sorry image, and thereby helped bring a new generation of bright young minds into the science.

Interest in dinosaurs, among scientists and public alike, was kept high by two other areas of research that burgeoned as the hot-blooded controversy ended.

In 1980, University of California/Berkeley physicist and Nobel-laureate Luis Alvarez and his geologist son, Walter, postulated that the collision of an extraterrestrial object with Earth, and a Nuclear Winter—like aftermath of the catastrophe, caused a mass extinction that killed off the dinosaurs and many other organisms 65 million years ago. Their evidence was the uncommon presence of the rare element iridium in rocks of that age worldwide. Only volcanic explosions and extraterrestrial impacts bring iridium to the earth's surface.

In the ensuing decade no natural-science story received as much media attention as did the dinosaur extinction theory. A host of new wrinkles, in theory and data, were added to the Alvarezes' theory. Asteroid showers might be the culprit, or else meteors, caused by the orbit of a "Death Star" and triggering, in turn, volcanic explosions. The iridium anomaly has been well documented and dated to very near the presumed date of mass extinctions at the end of dinosaur times. Changes in the numbers and types of plants, revealed in fossil leaves and pollen, provided intriguing new evidence that something drastic did occur 65 million years ago. In 1991, satellite imagery revealed what appears to be the "smoking gun," a crater, off the Caribbean coast of Mexico, big enough (120 miles in diameter), and old enough (65 million years), to have created drastic climatic changes wiping out the dinosaurs and many other life-forms.

Dinosaur paleontologists never embraced the impact-extinction theory wholeheartedly. Luis Alvarez had dismissed paleontologists as "stamp collectors." But paleontologists questioned the Alvarezes' theory strongly when it was first announced, pointing to the evidence for a more gradual but significant change in climate, geography, and dinosaur diversity before the proposed extraterrestrial calamity. Through millions of years the seas had been shrinking, the temperatures becoming more extreme, the numbers and kinds of dinosaurs (at least those we can find) declining.

Excavations by paleontologists around the world, and those of their predecessors, reveal no dinosaurs anywhere within one hundred thousand years of the proposed impact, while some controversial finds show dinosaur fossils from sediments laid down as much as one million years later than the suggested time of a big bang from space.

Yet suggestive new evidence from a variety of sources has caused many dinosaur paleontologists to acknowledge that a catastrophe of extraterrestrial causes may well have occurred. Climates now appear to have changed radically, not slowly, near the end of dinosaur time. And dinosaur diversity may not have declined slowly either. The quest for a deus ex machina to account for dinosaur extinction will go on, but physicists and geologists, not dinosaur paleontologists, will do the looking.

Both the hot-bloodedness and sudden-extinction controversies thrust dinosaur paleontologists into a defensive role—parrying attacks, first of Bakker, the renegade in their ranks, and then of Alvarez, the vituperative and celebrated outsider. While both controversies awakened public interest in dinosaurs, neither was key to the true reinvigoration of dinosaur science.

The Good Mother and the Drowned Children

More emblematic of the vigor of dinosaur paleontologists' own research—into how dinosaurs lived—was a discovery made two years prior to the extinction hypothesis, in 1978. That summer, Jack Horner and his friend Robert Makela discovered the nests, eggs, and hatchling babies of duckbilled dinosaurs in western Montana.

They named these creatures *Maiasaura* ("good-mother lizard") for the evidence they found, including regurgitated food and well-trampled nests, that suggested these dinosaurs cared for their helpless young well after birth. Such sophisticated, birdlike behavior is unknown in living reptiles.

In the same locality, Horner found the nests, eggs, and embryonic young of a smaller plant-eating dinosaur. Long-legged and smaller than an adult human, these dinosaurs pursued quite a different developmental strategy. Fossils from different growth stages of their young showed that these dinosaurs, like the *Maiasaura,* were fast-growing animals. But smoothly finished surfaces on the little bone ends, unlike the soft, pitted tips of baby maiasaur bones, indicated these animals were born up-and-running, capable of independent life immediately after birth.

Horner's spectacular discoveries are central to the fresh and ongoing investigations into the life cycle of the individual dinosaur. Baby dinosaurs were all but unknown before Horner's *Maiasaura* find. Though inspired in part by an amateur's discovery of baby dinosaur bones, Horner made his discoveries because he had searched not only in ancient river deltas and stream channels where dinosaur bones are most often found, but in formerly dry inland areas where many dinosaurs once nested.

Since his initial discoveries, Horner and his associates have found hundreds of dinosaur eggs, tens of nests, many babies of all sizes, and even a bone bed with an estimated ten thousand individuals of a migratory herd of *Maiasaura.* In Montana, Horner has found several varieties of dinosaur embryos, as has Phil Currie, just across the Alberta border. Dinosaur eggs, nests, and young have also lately been found in China, Mongolia, Transylvania, Argentina, India, and South Africa. These discoveries offer a raft of insights into what was until lately an unaddressed issue—how dinosaurs developed.

In Dinosaur Provincial Park in 1977, Currie began his discoveries of how herding dinosaurs lived and died. Exploring Dinosaur Provincial Park, he came upon a bone bed, a jumble of dinosaur fossils more than twenty-five thousand square feet in area. Nearly all of the bones Currie recognized as remains of horned dinosaurs of the genus "Centrosaurus" (now known as *Eucentrosaurus*). These were twenty-foot-long, single-horned cousins of the familiar *Triceratops.* The bone bed proved to be one of the largest concentrations of dinosaur bones known in the world, and perhaps the more revealing for what it shows of dinosaur life and death. In his taphonomic study, Currie determined that these bones were jumbled in the aftermath of one catastrophe. He surmised that the animals were herding, attempting to cross a flood-swollen river. Jostling in panic, they pulled one another down, much as do caribou in contemporary Canadian floods, or wildebeest in Africa.

The bodies drifted downstream, washed up on the river's edge, and were further exposed as the floodwaters receded. Fierce dinosaur predators ripped the meat from the bones, leaving their own broken serrated teeth among the bite-marked bones of their prey.

Predators also trampled the bones of the centrosaurs while they were still fresh, not fossilized, making distinctive spiral "green-stick" fractures. In the next flood season the disarticulated bones were scrambled and buried in a heap by another cycle of flooding.

The range of sizes of the horned dinosaurs Currie found at the site says much about how these dinosaurs lived and died. Most are young adults, suggesting that the catastrophe wiped out the herd indiscriminately. The range and distribution of sizes—three tiny centrosaurs, five more that were double that size, and the rest at least twice bigger yet—is consistent with the growth patterns of animals that breed seasonally. And such a growth pattern jibes with Dodson's analysis of the seasonal Canadian climate in these dinosaurs' time.

Further excavation and study, Currie observed, would make possible growth-rate calculations for young centrosaurs and conclusions about the physiology of horned dinosaurs. Work is not finished at the site, but the findings continue to suggest that at least some dinosaurs traveled in herds, that they visited the same rearing grounds each year, and that they migrated long distances.

With Currie's discoveries fresh in mind, and its pockets flush with oil revenues in 1978, the Albertan provincial government mandated the construction of the Tyrrell Museum of Paleontology in Drumheller. And it was at this magnificent facility, set snugly into striped badlands, that the world's dinosaur experts convened in 1988.

The Bone People

The Royal Tyrrell Museum is a graceful, low-slung structure, highlighted by plaza fountains. Within, more than eight hundred original fossils and thirty fiberglass skeletons and recreations are arranged in a wending tour of prehistory, against a backdrop of nearly five hundred feet of landscape murals. Half a million visitors journey to this isolated outpost each year, drawn by the allure of its dinosaurs.

Outside, more dinosaurs await admission. During construction of the museum, within a mile radius of the site, Currie and museum staffers found half a dozen major articulated dinosaur skeletons, including three large predators known as albertosaurs, a mummified duckbilled dinosaur, and a club-tailed armored dinosaur. Dinosaur skeletons continue to be unearthed from the area at a rate of one half dozen a year, four times as fast as they can be prepared by a staff of twenty preparators, the largest in the Western world.

The dinosaur experts who flocked to Drumheller in 1988, many driving across the continent for up to eighteen hours, came to the sleepy city not to dig in the badlands nor to just barhop, eat, sleep, and wince at the grotesque plaster dinosaurs on the occasional greensward.

For fifty years paleontologists who study backboned animals have gathered each fall for the annual meetings of the Society of Vertebrate Paleontology. SVP numbers 1,200 North American professional and amateur fossil enthusiasts, the many who study mammals and the far fewer who specialize in fishes, reptiles, birds, or dinosaurs.

As with most scientific disciplines, the ranks of paleontologists are filled largely by opinionated middle-aged men, though these researchers are, for the most, younger, less formal, and less vituperative than many other academics.

The dinosaur contingent, however, has more than its fair share of bickering and grandstanding. "It's getting to be like the guys who work on human origins," observes one outspoken mammal paleontologist, Spencer Lucas of the New Mexico Museum of Natural History. Acolytes feud with their former avatars, and little-known academics from hole-in-the-wall colleges vie against those from the dominating big-name and big-money institutions. Mediagenic personalities are accused of short-circuiting established research and publication procedures.

None of this internecine rivalry is peculiar to dinosaur science, but the spotlight of public attention shown on dinosaur researchers heightens their disagreements. The public cares about dinosaurs and little else in paleontology. "Nobody asks those of us who work on Permian reptiles to speculate on their social behavior," says Lucas.

Even at this scientific gathering, dinosaurs are a vogue. At the 1991 SVP meetings in San Diego, coincident with a cover story in *Newsweek* on dinosaur discoveries, Lucas recalled, "We used to get a few dinosaur papers at these meetings, but since Drumheller, more and more of SVP, up to one-third now, is devoted to dinosaur talks." Many of the dinosaur presentations, to Lucas's mind, are weak: "There are a lot of untrained people who do things with dinosaurs just because everyone loves dinosaurs, creating a perception among other paleontologists that dinosaur science is for media hounds, not serious scientists. That view is partly personal and professional jealousy and partly serious concern about what is useful science. Dodson, Currie, Ostrom, and some others do careful work, but there are a lot of dinosaur people who don't."

The bad work and bad feelings did not predominate at the Drumheller conference. VPs spent the weekend swapping stories and fossils in corridors, preparation laboratories, and meeting rooms. And principally, they clustered for a series of rapid-fire formal presentations of new and as yet-unpublished discoveries, theories, and techniques.

Dinosaurs were but one item on the fossil smorgasbord at Tyrrell's 1988 convention, one afternoon's worth of formal presentations. Yet more than any other conference of recent years, this gathering was dedicated to celebrating both the science's history and its unprecedented current vigor, in the locale where so much of both is centered. With the exception of Ostrom, North America's leading dinosaur experts and personalities were on hand, from the shy, avuncular specialist in giant dinosaurs, Jack McIntosh, to the self-declared maverick, Bob Bakker, his ponytail cascading from his Stetson.

The Tyrrell SVP convention provided the first opportunity for dinosaur diggers to learn of a host of outstanding finds made that summer, including those by the museum's chief paleontologist, Currie, and others on the Chinese-Canadian expedition to the Gobi Desert, which the museum helped sponsor. Those finds included dinosaur eggs, baby armored dinosaurs, and the largest dinosaur in Asia (as described in Chapter 9). Most of these finds have yet to be presented to the public, as it customarily takes several years to complete laboratory preparation and analysis and publish scientific papers announcing a discovery.

Several of the other fresh discoveries were spectacular reminders of the prodigious size of some dinosaurs, fossils of staggering proportions. Milwaukee Public Museum volunteers had unearthed a horned dinosaur skull in the Hell Creek Formation of eastern Montana. More than ten feet wide by eight feet high, it is the largest skull of any land animal ever found. A Brigham Young University team had unearthed a dinosaur pelvis that was six feet high and thought to be the biggest single bone yet found. Utah state paleontologist David Gillette described the ongoing excavation of his New Mexico *Seismosaurus*, the biggest, he believes, of all known dinosaurs (see Chapter 7).

And when a talented investigator looks in the right environment, he can uncover dinosaurs in wondrous plenty. So Jack Horner had done that summer and many others. The laconic Montanan gave a few details of "Jack's Birthday Site"—a western-Montana hillside he'd ambled across on his birthday in June that was so densely packed with remains of several dinosaur species that the entire slope glistened in the morning light.

There were other causes for celebration among dinosaur paleontologists at the SVP meeting. The growing sophistication of dinosaur studies, long lightly regarded even among the other "soft" sciences, was recognized via National Science Foundation grants to two of the discipline's brightest youngest theorists, Jacques Gauthier and Paul Sereno, to study dinosaur origins.

Unlike the contentious meetings of the 1970s, this conference and those since the early 1980s have not been dominated by acrimonious debate over whether dinosaurs were warm- or cold-blooded, and whether they died out suddenly or gradually. In a summary of a 1986 Tyrrell dinosaur conference, Dodson noted "a poll was conducted on the tepid issues of our day. On the subject of the tempo of dinosaur extinction: . . . thirty-eight votes for gradual extinction, four votes for catastrophic extinction, one undecided. Somewhat more interesting was the opinion on the metabolic level of small theropods [meat-eating dinosaurs]: thirty-four thought they were endothermic ["warm-blooded"], six thought they were ectothermic ["cold-blooded"], and three were undecided."

Instead of controversy, in Tyrrell conference rooms, for the following three days, the dinosaur researchers sampled from collegial presentations of ongoing research. No other talk excited the listeners as did Paul Sereno's. Sereno regaled the crowd with slides of his explorations in the Ischigualasto Formation in northeast Argentina.

The scenery was dramatic, an exposure of rock face so high and precipitous it was breathtaking, even in slides. But what elicited many audible gasps from the assembled paleontologists was a slide of a foot-long curved series of red-black bones. This, discernible, even in its unprepared state, was the perfectly preserved skull and neck of a dinosaur.

This was Sereno's latest find, the first known remains of such parts from perhaps the earliest-known dinosaur of them all, and a key clue in solving one of the great puzzles of dinosaur science—how, when, and where did dinosaurs originate.

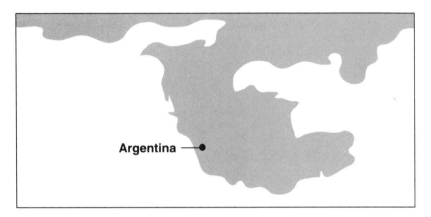

Pangaea, 231–225m. yrs. ago

225m. first dinosaur   65m.

AGE OF THE DINOSAURS

First animals   First land animals   First human

590m.   350m.   245m. 208m.   145m.   65m.   Today

CHAPTER 2

The Search for the First Dinosaur

Something wonderful was going on around the world some 225 million years ago, an era unusually rich in large and spectacular animals. It was a time for grand experiments, when pterosaurs became the first animals to conquer the air, and dinosaurs began to walk the earth.

Large reptiles had already returned to the sea, establishing their dominance as ichthyosaurs and plesiosaurs. Sharks swam through salty, shallow oceans, and lobe-finned fish coursed through rivers and lakes. In search of these and other prey, huge pterosaurs, gliding reptiles, and possibly the ancestors of modern birds took to the air.

All lands were joined in a single supercontinent. Warm climes supported a rich flora of spore-bearing ferns, club mosses and horsetails, primitive seed plants, cycads, early ginkgoes and conifers spreading from the banks of seasonally engorged rivers. In moist forests ferns carpeted the understory, deeply shaded by a two-hundred-foot-high crown of uniform, branchless trees, spaced but ten feet apart.

Though linked, the verdant lands north and south of the equator seem to have had separate monsoon-dominated climates and distinct plant communities. Whatever the conditions, animal life on land responded by flourishing as never before. Rather suddenly, from the evolutionary perspective, crocodiles, turtles, lizards, frogs, and mammals all appeared for the first time in the fossil record. They coexisted with holdovers from the Permian period in a spectacular menagerie.

The biggest creatures along lush, reed-choked streambanks were armor-plated lizards up to thirty feet long, scarfing up ferns and thick-stemmed palmlike plants in their wide muzzles. And around their feet scampered some of the smallest creatures of the day, the ancestors of modern mammals. Among the other giant browsers were "cow-turtles" as large as small cattle, with big, toothless beaks. Equally massive were rhynchosaurs, lizards equipped with powerful, clawed hind limbs to dig roots and broad owlish faces packed with the muscles needed to chew their food. Still more stout lizards, these the size of pigs, used their powerful teeth to consume a varied diet of plant leftovers. All lived in fear of huge and savage predators, half-wolf, half-crocodile in appearance, measuring twelve to twenty feet from head to tail.

Into this world came yet another predator, unlike any of the others, and one of the first of a kind that would in several million years time come to dominate the landscape, in a host of forms, for over 100 million years. This creature, three feet in length, stood tall on hind legs, flashing sharp teeth in a powerful, gaping jaw. It was among the very first of the dinosaurs.

In March of 1988, a team of ten Argentine and American paleontologists entered this prehistoric world, or rather what remains of it, in the stark badlands of northwestern Argentina, two hours' drive by jeep jalopy from the nearest town. Their leader was Paul Sereno, a brilliant young American paleontologist; their host, Jose Bonaparte, perhaps the world's most accomplished dinosaur-digger. Their grail was fossil evidence of a dinosaur more than 225 million years old—the earliest dinosaur yet known.

*Youthful Paul Sereno is already among the world's most traveled
and accomplished dinosaur paleontologists. Here in his University
of Chicago laboratory, he holds the skull of a psittacosaur, a small,
parrot-beaked dinosaur. Sereno has identified psittacosaur species
from China. [Photo by author]*

Jose Bonaparte, "Master of the Mesozoic," is South America's leading dinosaur and early mammal expert. He's pictured here in Ischigualasto, Argentina, where he made many dinosaur discoveries. [Photo by Paul Sereno]

The "first" dinosaur can never be found, as Sereno well knew, just as there is no single "missing link" in human evolution. With luck and enterprise, we can always probe farther back in time, discovering creatures with increasingly intermediate features between dinosaurs and their reptilian ancestors. Nor, if we set an arbitrary standard for the evolution of various features used to define some animals as truly dinosaurian, can we be satisfied that there is not, hidden somewhere, an older fossil than any we possess that also satisfies our prerequisities for what makes a dinosaur.

Still, locating the earliest dinosaur yet known was a scientific endeavor that might garner far more than headlines. Sereno hoped it might help resolve the many questions about the evolutionary origins of dinosaurs, questions that have vexed paleontologists and muddled systematics for decades. While physicists have devoted considerable thought to the question of dinosaur extinction, dinosaur paleontologists have concerned themselves more with where dinosaurs came from. "Origins are the neat stuff of evolution," waxes University of California/Berkeley paleontologist Kevin Padian. In the origin of dinosaurs, or any group of animals, rests not only another chapter in life's history, but the key to defining the special identity of that group. Neat stuff indeed.

As recently as 1984, the prevailing scientific view was that dinosaurs were an unnaturally grouped collection of animals. This hodgepodge belonged to at least two lineages that arose from several different ancestors some 225 million years ago. Only lately, through recent discoveries and through new techniques for analyzing and grouping fossils, has a different view emerged, of dinosaurs as varied creatures descended from a common rootstock. Sereno himself was among those who had pioneered the application of new classification techniques to dinosaurs. So he came to the Argentine badlands with a single locality, and a single early dinosaur, in mind.

At the foot of the Andes, Sereno's team came to an enormous, barren depression running north-south and strewn with sandstone towers, mesas, and huge rust-red lumps of ironstone. This forbidding land, known on tourist maps as the Valle de la Luna (Moon Valley) or more properly as the Ischigualasto Valley, lies largely within a provincial park, a refuge for the wildlife of the region—including guanaco, rheas, and armadillos. Bordering the valley on one flank is a nearly mile-high wall of cliffs. To the world-traveling Sereno it was the most spectacular terrain he'd ever seen—snowcapped mountains in the distance, and around him shapes and colors to rival the Painted Desert of the American West.

The immense badlands of Ischigualasto, in northwestern Argentina, are home to the world's oldest known dinosaur fossils. [Photo by author]

Sereno, Bonaparte, and their party had come to Ischigualasto because the ironstone of the valley was known to be the resting place of thousands of fossils. The sediments of the Ischigualasto Formation are 220 million years old, and two even older rock layers were deposited as long as 228 million years ago. The striped-rock exposures of Ischigualasto age are themselves enormous—fifty miles long and nine hundred feet thick, and full of fossils. The site is one of the longest continuous stretches of Triassic geology open to examination anywhere on earth.

The Ischigualasto Valley has long been known to contain important dinosaur fossils. In 1930, German paleontologist Friedrich von Huene visited on muleback, on a side trip from exploring other equally ancient deposits in southern Brazil. He recommended an Ischigualasto exploration to famed Harvard mammal paleontologist Alfred Romer, who belatedly mounted a dig in 1958. Romer returned in 1964 to a site fifty miles to the northeast of the valley and discovered more fossils. (Some of these treasures were impounded by overzealous provincial police, others removed to Harvard by Romer's assistant, and later the nation's premier dinosaur-finder, Jim Jensen, via a wild Andean truck ride, police in hot pursuit.)

Sereno's host on the 1988 expedition, Jose Napoleon Bonaparte, was himself a protégé of Romer's who first came to Ischigualasto in 1963. As a self-taught paleozoologist, keeper of the fossil collections of the University of Tucumán, and now chief of vertebrate paleontology at the Argentine Museum of Natural Sciences, Bonaparte has nearly single-handedly changed our understanding of South American life in dinosaur times. No country, with the possible exceptions of China and the United States, contains such a broad range of dinosaur fossils across the full span of the creatures' existence as Argentina.

Balding, bespectacled, and nearing sixty, Bonaparte is a deceptively bookish-looking man. Vigorous and strong-willed, he still leads expeditions in search of Mesozoic mammals and dinosaurs each summer. From sediments laid down after the Triassic, when the dinosaur world split in two, Bonaparte has documented a variety of peculiar animals that testify to South America's isolation through much of dinosaur time. He has excavated huge armor-plated, plant-eating dinosaurs, and other herbivorous dinosaurs with peculiar double sail fins, as well as Roman-nosed and odd short-limbed large predatory dinosaurs and hook-toed small ones, and several distinctive forms of crocodiles, birds, and mammals. All were absent from the northern hemisphere. Bonaparte has also discovered some of the earliest-known dinosaurs and their reptilian ancestors in the Triassic rocks of northwestern Argentina.

Another of Bonaparte's former mentors, Argentine paleontologist

Herrerasaurus *[John Sibbick]*

Osvaldo Reig, identified three new genera of dinosaurs from skeletal fragments found in Ischigualasto, a small dinosaur, a large one, and an intermediate in size that he dubbed *Herrerasaurus ischigualastensis,* its genus name honoring a local rancher and guide, Don Victorino Herrera.

From fragmentary remains it was evident *Herrerasaurus* was four-toed and unusually short in the thigh for a dinosaur, both apparently primitive features. The age of the Ischigualasto rock, only loosely estimated from the various animals fossilized within, indicated that *Herrerasaurus,* whatever it was, was among the earliest dinosaurs known.

More recently, other apparently earlier dinosaurs have emerged. One was "Gertie," a much-heralded 1985 find from the Petrified Forest in Arizona, which turned out to be far less substantial and not as old as originally thought. The most viable candidate for oldest known dinosaur has been *Staurikosaurus,* an agile, small, predatory dinosaur less than seven feet long and weighing seventy pounds. This "Cross lizard" (for the Southern Cross constellation) was found in southern Brazil by ex–American Museum of Natural History paleontologist Edwin Colbert, who named it in 1970, following a lead from the peripatetic German explorer Von Huene. The rock strata from which *Staurikosaurus* emerged was believed to be 230 million years old.

Subsequent studies demonstrated that *Staurikosaurus* came from rock no older than that in which *Herrerasaurus* was buried. Still, the role of *Herrerasaurus,* or any early dinosaur in dinosaur evolution, could not be fully appreciated until all dinosaurs and the shape of early dinosaur evolution were better understood. In that prodigious task, Paul Sereno, barely into his thirties, has played a major role. Boldly adventuresome, boyishly handsome, impassioned yet diligent and insightful in his research, Sereno wears the mantle of a scientific pioneer easily.

One of six children, all of them academically gifted, of an Illinois mailman and a housewife, Paul Sereno planned on pursuing his artistic talents until he caught the fossil-hunting bug on an undergraduate visit to the American Museum of Natural History. He searched for Ice Age tortoises on Howe Island, Australia, and traveled around the world, with stops in China, Mongolia, and Siberia, as an American Museum graduate student.

Sereno discovered dinosaurs at the American Museum "not as strange monsters, but as a huge group of animals desperately in need of putting in some scientific order." And with the world's largest collection of dinosaurs at his fingertips, nothing else was needed to begin the reorganization "except to travel widely, gathering additional data on a global scale."

His first dinosaur travels were his most extensive. Financed by government grants, he became in 1984 the first American paleontology student in decades to study in China. With $10,000 in cash tucked under his vest in a holster, he bought a ticket to Beijing and a return flight from London—eight months later. In that time he would log twelve thousand miles via Mongolia, Soviet Siberia, and Europe. Through sheer persistence, he became the first Western scientist in nearly a half century to visit the spectacular Flaming Cliffs of Outer Mongolia where the first dinosaur eggs and nests were discovered.

In the spring of 1990, Sereno made an ambitious jaunt—overland (with the occasional ferry) from England through the Sahara to Niger. With British Museum researcher David Ward, Sereno was following up on the explorations of a British team four years before, and the French two decades earlier. Those treks had produced promising chunks of huge brontosaurlike dinosaurs in the Sahara from a little-known time 110 million years ago, early in the last dinosaur era, the Cretaceous.

Following maddeningly inaccurate maps left by the French explorers, Sereno and his colleagues did find several promising sites, including one from near the end of dinosaur time. But a chance stop at the office of a local chieftain led him to his richest discovery. Sereno noted a chunk of dinosaur bone on the man's desk. "He said, 'Oh, that thing! I can show you lots of those!' "

There followed a winding journey of several hours in which the caravan and its guide lost the way several times. At last, Sereno stopped in frustration, while the guide continued in another vehicle. "Within a half hour they were back, yelling that they'd found it."

"It," Sereno soon discovered, was a vast graveyard of giant dinosaurs, a plain of mound after mound of slightly jumbled behemoth bones. "Each pile seemed to belong to a different dinosaur. It was really bizarre," Sereno recalls. From the soft dirt, Sereno was able to exhume a femur, a leg bone longer and wider than his nearly six-foot-tall frame. Chopped in four sections, it was transported to London, where it sits today in David Ward's study, awaiting reunion with the rest of the dinosaur's more than seventy-foot-long frame after Sereno resumes his Saharan adventures in 1992.

In the Sahara Desert in Niger, Paul Sereno uncovers a leg bone of a titanosaur. [Photo courtesy Paul Sereno]

As in Ischigualasto and the Sahara, Sereno has demonstrated a knack for finding significant fossils in all his far-flung travels. "It must mean I'll die young," says Sereno of his inexplicable good fortune with fossils. On a one-day student outing to Big Bend National Park in Texas in early 1991, Sereno's charges came upon the first complete skull of a huge horned dinosaur, *Chasmosaurus*—a find that had eluded dinosaur-diggers in that well-explored region for nearly a century.

The wonder of these discoveries is never lost on Sereno, as he readily admitted to a capacity lecture audience at the Field Museum in early 1990. "The Creationists will kill me for this, but we are playing God. We get to create new species. It's an incredible thrill!"

Thrill and adventure notwithstanding, it is the daunting business of linking species into a comprehensive and comprehensible framework that consumes most of Sereno's time. When Sereno began his peripatetic studies, an emerging theoretical tool called cladistics was splitting the ranks of dinosaur scientists and sparking new inquiries into dinosaur origins. As an elegant means of classifying organisms, living and extinct, cladistics fired Sereno's intellectual interest, and he sought in his travels, and in detailed papers that emerged from them in 1984 and 1986, to apply this nascent discipline to the muddle of dinosaur classification.

In the Beginning

The origin of the dinosaurs, and ourselves, can ultimately be traced, however imperfectly, back to the first fossils of blue-green algaelike organisms in rocks 3.5 billion years old. The Carboniferous, 345 to 280 million years ago, however, seems a better place to start, for by then the evolutionary path of our ancestors and the dinosaurs' had already split.

It was then that the first reptiles arose from amphibian ancestors, heading off in two evolutionary directions. Some evolved skulls with a single large opening behind each eye socket. The first of these were large sprawling animals, including the sail-finned *Dimetrodon*, commonly mistaken for a dinosaur. Rather, it is one of our ancestors, for mammals evolved from this single-skull-opening stock. Even earlier, rat- to wolf-sized carnivores and plant-eaters the size of small hippos began to develop, the animals that ruled the land through the greatest extinction in the history of life—at the end of the Permian era, some 230 million years ago. No doubt that extinction was fostered by the serendipitous coalescing of all the land masses of the planet in the Permian and their northward shift. The massing of land destroyed productive shallow coastal margins, and the poleward movement may have led to a devastating ice age.

Yet, by modern standards, the climate that followed was warm and dry with monsoon storm systems, a regime that would prevail for nearly 200 million years, well into the Cretaceous, the last dinosaur period. The single-skull-holed therapsids ruled on through the Permian into the next era, the Mesozoic, where some of these animals— tiny, shrew-sized, and perhaps furred—developed a new jaw joint. The evolution of this feature is what made them, by definition, mammals. (Other characteristics by which we identify modern mammals, among them hair and mammary glands, are not detectable in fossils.)

Meanwhile other reptiles—the ancestors of dinosaurs—were evolving with two holes on each side of the skulls. In another sudden widespread appearance in the fossil record, the two-holers evolved from lizardy creatures scarcely a foot long into the first carnivorous archosaurs ("the ruling reptiles"), including the ancestors of dinosaurs. The archosaurs had branched off early in the Triassic (some 240 million years ago) into several lineages, all tending toward a more upright posture than earlier reptiles. All looked at least remotely, sometimes strikingly, like modern crocodiles. With anatomical modifications, such as the curving of their leg bones, they were not just more mobile, but able to attain larger size. Their increased weight could rest on the sturdy legs, instead of pressing a dragging belly flat to the ground.

But further anatomical changes would be necessary for the fully upright walking and running characteristic of dinosaurs. And somewhere between these predinosaur archosaurs and *Herrerasaurus,* these anatomical features evolved. Two of the most significant alterations involve the hip and the ankle. In most semierect archosaurs two rectangular bones in the ankle were joined by a peg and socket. But in dinosaurs (and birds and pterosaurs) peg and socket are gone. (See page 24.) The ankle bones are firmly attached to the shin bones. This new ankle structure provided the greater strength and stability necessary to support dinosaurian bulk.

Another marked anatomical restructuring is the dinosaurian hip. Imagine, for a moment, your own thighbone. Or picture a chicken drumstick. Dinosaurs, like birds and us, had a ball-topped thighbone that inserts into the hip socket. In semi-upright archosaurs a "headless" femur juts at an inward angle into the shallow-grooved hip socket. But in dinosaurs the femur runs straight up and down. At its tip an inwardly projecting head with a bulbous end fits into a lip on the socket of the hipbone, a bone known as the ilium. An opening has appeared in the socket between the ilium and the two other pelvic bones.

And early in the dinosaurs' evolution, or sometime before the first true dinosaur appeared, as it was long believed, dinosaurian hip structure evolved in two distinct directions. One is the "lizard-hipped" line of dinosaurs, which numbers huge vegetarians such as *Diplodocus* and meat-eaters such as *Tyrannosaurus rex*. These dinosaurs all possessed the same pelvic structure, with three hip bones that radiated in different directions. In the "bird-hipped" line, the third of these hip bones is turned parallel to the second (see diagram, page 25). Both alignments appear to have been capable of supporting dinosaurs of hugely varied sizes and speeds.

These innovations in ankle and hip architecture characterize the development of something big, fast, and two- or four-legged: a dinosaur. But the question remains—what exactly was a dinosaur? That definition has never been clearly drawn. Well above the level of species—populations that can interbreed—organisms have traditionally been grouped according to one or more often dramatic characteristics that appear to set them apart from other animals. Birds are birds because they alone have feathers. Dinosaurs were dinosaurs because they lived on land in the Mesozoic era, or because they were big reptiles that stood erect. Sometimes dinosaurs were defined instead by what they lacked—feathers and fur.

To generations of paleontologists, dinosaurs were not much more clearly distinguished. Yes, some significant anatomical characteristics, particularly in the hip and pelvis, set them apart from their land-animal contemporaries and predecessors. But ankle and hip structure also united them with birds, and hip structure divided them in two: bird-hipped and lizard-hipped.

For decades, new dinosaurs have been found, described, and categorized within the bird-hipped or lizard-hipped lines. Cladistics, however, groups organisms not by their outstanding differences or their nearest ancestor, but by the many, often subtle, derived (new) characteristics they shared. Cladistics was developed to make sense of the frightening variety of living beetles by a German entomologist, Willi Hennig, in 1950, in a book not translated into English until 1966. According to Hennig's theory, categories called clades could be organized around a group of organisms, which would include an ancestor and all its known descendants. The relationships between these groups are indicated by neatly branching diagrams, cladograms. The determining characteristics of a clade must be shared features that are advanced in form. Primitive features are not included, as these characteristics might have arisen prior to the origin of the group. For example, a five-toed foot is a primitive characteristic, as it is shared by humans and many living salamanders. Clearly, these groups are not closely related.

What must be used to group organisms, say cladists, are synapomorphies, or shared derived characters. These features are evolutionary novelties. They point to the recent common ancestry of those species that share those novelties. For example, feathers on birds are synapomorphies, a derived feature for birds. But the laying of eggs is not a synapomorphy of birds. Crocodiles and snakes also lay eggs.

In drawing patterns of relationship, cladists face the thorny problem of choosing characters that are truly derived, winnowing out those that are shared by convergence or parallel evolution. For example, the panda has a thumb, but it has developed one from its wrist by its own peculiar evolutionary pathway, as Stephen Jay Gould has pointed out. Finding a proper derived character is not a task easily done. In a famous example of the confusion cladograms can foster, consider that they can make cows and lungfish appear more closely related than lungfish and salmon. That happens because the ancestors of terrestrial vertebrates, cows included, branched off from the lungfish group after both had branched from the bony fishes, salmon among them.

Scientific critics have attacked cladistics as a technique masquerading as philosophy, a pseudoscience full of jargon—even, in one venomous attack upon a cladistically inspired British Museum exhibit, as Marxist dogma. To this day some palentologists, Bonaparte included, have no interest in cladistic analyses of their discoveries. Yet cladism has been put to increasing scientific use since the early 1970s. Despite the confusion it has caused its critics, cladistics has helped in defining the little-understood and much-argued relationships among dinosaurs. And cladistic studies, by Sereno and others, have helped identify features that distinguish dinosaurs as a legitimately distinct group, features found in a single ancestor.

Sereno and others began redrawing the lines of relationship among dinosaurs and other animals cladistically in papers first published in 1984. Their method calls for a thorough examination of all relevant dinosaur fossils, preferably firsthand, recording all significant anatomical features. Then, with the aid of a computer (because the list of characters may run to several hundred over a whole skeleton), a cladogram is drawn, showing the lines of relationship among animals sharing the greatest number of these features.

In this laborious manner, Sereno reexamined the bird-hipped dinosaurs. He made new sense of the evolutionary relationships between its well-known subgroups—duckbills, stegosaurs, armored dinosaurs, dome-headed dinosaurs, and horned dinosaurs. And Sereno began comparing all dinosaurs to their nearest nondinosaurian relatives, in a further consideration of the features uniquely shared by dinosaurs. California Academy of Sciences herpetologist and paleontologist Jacques Gauthier did the same when he subjected the lizard-hipped dinosaurs to cladistic study.

Gauthier compiled a list of nine characteristics (synapomorphies) that unite all dinosaurs, features that also set the dinosaurs apart from other archosaurs: among them the dinosaurian shape of the hand, foot, thighbone, and "drumstick" ankle. In a 1986 paper, Gauthier went so far as to reason what the ancestral dinosaur, if one could be found, would look like: it would be bipedal, small (less than seven feet long), with short forelimbs better suited for grasping than locomotion, a short trunk, and hind limbs built for support and movement. It would have a hinge ankle and perforated pelvis. It would be three-toed, or nearly so, and capable of standing on those toes. It would have had a long stabilizing tail that moved up and down, and a mobile and S-shaped neck.

To probe for dinosaur origins, to perhaps find something close to this ideal ancestral dinosaur, there could be no better place than where Sereno went looking on his 1988 dig, Ischigualasto. *Herrerasaurus* might well qualify for a central position in dinosaur evolution, but to determine this more of *Herrerasaurus* was needed. A skull, often the most revealing of all body parts in determining an animal's evolutionary status, would be the best evidence of all. But most fossil dinosaur skulls are fragile and exceedingly hard to come by. Expeditions rarely succeed with such specific agendas. Dinosaur paleontology advances by slow and painstaking work in the laboratory and in the field, seldom by "Eureka!" moments of discovery. Rather, Sereno and colleagues planned to survey the region, gathering as much information as they could about all the fauna of one place in early dinosaur times. Finding the head of a *Herrerasaurus* was but a dream, an unspoken one, of Sereno's.

Gathering fresh evidence of the earliest dinosaurs was not, however, only Sereno's goal. "It was a plum waiting to be picked," as one fellow paleontologist put it. Gauthier, himself a young and accomplished researcher on dinosaur origins, was after the same sort of evidence, searching not in the field but in the museum collections of Triassic dinosaurs worldwide. But in quietly linking up with Bonaparte, and securing National Science Foundation funding for their joint expedition, Sereno had found his way to the best source of early dinosaurs, the Argentine badlands.

Bring Me the Head of *Herrerasaurus*

At Ischigualasto, the paleontologists and their Argentine–North American crew found plenty of fossils. They excavated remains of more than one hundred animals—primitive dinosaurs and a host of their contemporaries. One morning in the third week of the dig, Sereno insisted on visiting a corner of the great basin that the group appeared to have missed in its early prospecting. At the basin's edge, Sereno recalled, he "gazed at this little back valley from a lookout ridge" and daydreamed "that it was there that a whole early dinosaur lurked." His chance to research his hunch came on a day off, after a morning spent taking field-camp photos.

The world's oldest dinosaur skull. Paul Sereno discovered this skull of Herrerasaurus *in Argentina in 1988. [Photo courtesy Paul Sereno]*

After a long drive and hike with his crew, Sereno was the first to descend into the badlands. "I laid my pack on a prominent rocky spine and walked off fifty feet, straight to the most complete early dinosaur skeleton and the first herrerasaur skull ever discovered. At first I thought it might be just another bloody rhynchosaur [cow-turtle]. Then my eyes slowly stopped along what appeared to be elongated neck vertebrae—literally one by one—right up to the beautiful occiput [back of the skull] of a primitive dinosaur." The foot-long *Herrerasaurus* skull lay exposed, its fanged profile in open view, still articulated with the neck and part of one arm, in the midst of a sandstone ledge, stained black by the iron-bearing mineral hematite. The fossil was elegantly preserved, locked into place by sand grains that had washed out of rivers and around the bones when this land was well irrigated, millions of years ago. On seeing "such a beautiful skull," Sereno recalls, "I literally went crazy, yelling. The crew slowly trickled in, amazed at the sight. When they heard my yells, they either thought I had found just what I'd found or that someone had died. I returned to the skeleton after pacing some distance and wept in front of all the crew—it was all over, we had done it!"

When the exultation ended, the painstaking preparation and analysis began. Even without considering the new discovery, *Herrerasaurus* has some "oddball characteristics," as Sereno puts it. It was, as far as is known, the first of the big dinosaurs, one thousand to two thousand pounds, twelve to fifteen feet by Sereno's early estimates. Among its peculiarities for a dinosaur, *Herrerasaurus* apparently lacks the beginning of a gap between hip bones, the pubis and ischium. And it doesn't have a dinosaurian crest on top of its femur, only a bump. The backbone, says Sereno, is "downright weird. The neural spines come straight up [they are long and outwardly curving on other dinosaurs]. They're like little squares. No other dinosaur has anything like that." Such peculiarities put *Herrerasaurus* slightly off the main track of early dinosaur evolution, an early dinosaur that appeared, in some respects, to have gone its own peculiar evolutionary way.

Just where *Herrerasaurus* fits into early dinosaur evolution is as yet uncertain. Is it from the rootstock of primitive dinosaurs from which both bird-hipped and lizard-hipped dinosaurs diverged, or does it belong to one of those main dinosaur groups?

When preparation of the skull was completed by Argentine scientists, Sereno and colleagues found this *Herrerasaurus* so splendidly preserved that they could identify minute bones of the ear and a fossilized ring around the animal's eye. From the size of the skull, Sereno estimated the animal's length to be but six feet to eight feet, its weight five hundred pounds. Those numbers are roughly half of the early estimates of *Herrerasaurus*'s bulk, and closer to Gauthier's estimates for the size of a first dinosaur, but it may well represent a half-grown animal. In other respects *Herrerasaurus* matched Gauthier's speculation about the anatomy of a first dinosaur. What's more, this most complete of Ischigualasto dinosaurs revealed that the two other kinds of dinosaurs named from more fragmentary remains in the region were also herrerasaurs, one a juvenile, the other a full-grown adult version of the subadult Sereno had unearthed.

Sereno and his Argentine collaborators have noted features on their *Herrerasaurus* that make the animal seem in some ways like a composite of different kinds of dinosaurs, and in still others unlike both bird-hipped and lizard-hipped dinosaurs. In particular, the skull featured a double-hinged jaw. Pivoting both at the rear and in the middle, it allowed the predator to ensnare an animal too large to be swallowed whole. The double-hinged jaw is a feature not otherwise known among predatory dinosaurs for another 50 million to 100 million years. Its appearance in *Herrerasaurus* supports cladistic analyses of the animal as an evolutionary diversion, off the main line of predatory dinosaur development. Nevertheless, says Sereno, *"Herrerasaurus*'s status as the world's oldest and most primitive dinosaur seems secure." At least, that is, until more exploration is done. Earlier dinosaurs, and ancestors to dinosaurs, will be found, and they may well be found in and around Ischigualasto. For not only has Ischigualasto produced fragments of *Herrerasaurus,* but thirty miles away, along the provincial border, the best candidate yet for an ancestor to the entire dinosaur line was found.

Discovered in a formation that lies below Ischigualasto's sediments was a foot-long creature named *Lagosuchus* by Romer for the rabbitlike appearance of its slim, long-shinned leg bones. Romer himself located its partial remains and published the first papers on it in 1971 and 1972. Bonaparte unearthed a nearly complete skeleton in 1975, but without a skull.

Nonetheless, the remains contained ample clues that, as Bonaparte says unequivocally, "it is an ancestor, closely related to dinosaurs." *Lagosuchus* is, in several ways, an evolutionary intermediate. The hip cup holding the femur is almost closed, not open as in dinosaurs, nor closed as in previous archosaurs. And the ankle bones look, from front to back, like a dinosaur's. Yet from the side they resemble an earlier archosaur's. *Lagosuchus* still has a heel on one of its ankle bones, a

SIBBICK ©

Lagosuchus *[John Sibbick]*

An artist's skeletal restoration of a bizarre and as yet unnamed South American sauropod dinosaur with a mane of long paired bones. Jose Bonaparte will be naming the animal soon. [Photo courtesy Jose Bonaparte, Museo de Ciencias Naturale, Buenos Aires, Argentina]

remnant of the peg-and-socket joint from earlier archosaurs, and a feature abandoned in dinosaurs. By the traditional method of defining dinosaurs by one or two revealing features, the *Lagosuchus* ankle might be sufficient to place it as the dinosaur's nearest ancestor. For cladistic analysis more fossils of other telling body parts are needed. Cladists might say more about where exactly *Lagosuchus* stands in dinosaur evolution, "if only," as Gauthier said, "a skull, hand, and intact foot were found."

So, Sereno returned to northern Argentina in the fall of 1991 to search for *Lagosuchus*. He thinks he may have found parts of it (as well as a new small predatory dinosaur) among his fossil discoveries, and he's begun analysis of the new material. No longer is he restraining his hopes. "You can never bet on finding something, but I'd say it's fifty-fifty we've found it." Those are high odds indeed in the chancy game of fossil-hunting.

Head of Carnotaurus *by pre-eminent dinosaur sculptor Stephen Czerkas. [© Stephen Czerkas; photo courtesy Sylvia Czerkas]*

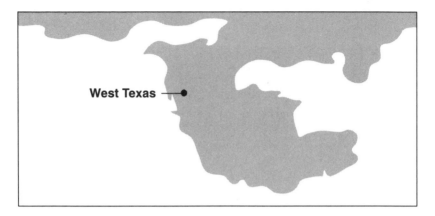

Pangaea, 225m. yrs. ago

225 m. yrs. ago

245m.   65m.

AGE OF THE DINOSAURS

Triassic Jurassic Cretaceous

First animals   First land animals   First human

590m.   350m.   245m. 208m. 145m.   65m. Today

The Case
of the Flying
Dinosaur

Look up in spring and see the Canada geese flying north. The dinosaurs are migrating once again. It is a compelling image—dinosaurs alive today as birds—popularized by the best-known of dinosaur paleontologists, Robert Bakker. And the notion of birds as dinosaurs fits the popular new image of dinosaurs as agile, lively, highly successful animals.

The flying-dinosaur image is not only appealing, it represents the now-dominant thinking among paleontologists that birds evolved from one branch of a widely diversified dinosaur family tree, at a time near the midpoint of the Dinosaur Age.

The dinosaur-bird connection is fashioned from many anatomical similarities, but precious few fossil links. But a new and controversial fossil discovery, at once strikingly birdlike and surprisingly ancient, threatens to force a paradigm shift in evolutionary thinking—the replacement of an established theory with an entirely new perspective.

Protoavis (top), Archaeopteryx *(bottom) [John Sibbick]*

The first dinosaurs appeared some 225 million years ago in the Late Triassic, when the world held but one continent, the monsoon-swept, subtropical Pangaea. The first feathered creature we know, *Archaeopteryx,* appeared in a very different and far more recent world—the Late Jurassic period, 145 million years ago. This was the time of the giant sauropods, the largest animals ever to walk the earth—among them *Diplodocus, Apatosaurus,* 140-foot-long *Supersaurus,* and six-story-high *Ultrasaurus.* The sauropods roamed semiarid lands on two super-continents that were in the process of separating into several more.

At that time, birds developed from the small carnivorous dinosaurs known as theropods, or so most paleontologists now think. For a century *Archaeopteryx* has been the consensus candidate for the oldest ancestral bird—close kin of dinosaurs and the most valuable of all fossils.

In fact, *Archaeopteryx* so closely resembles some of these theropods that one of the six known specimens of *Archaeopteryx* was, for several decades, mistakenly classified as the theropod dinosaur *Compsognathus.* Its proper identity was not discerned until 1985 when a scientist detected a wispy outline of feathers on the fossil.

Just where birds came from, evolutionarily, has troubled scientists for more than 130 years. The origin of birds ranks as one of the most dramatic transformations in the history of life. The scales on birds' legs and the fact that they lay hard-shelled eggs suggest to the casual observer that birds descended from reptiles. But which reptiles, and when? If, as a minority of paleontologists now think, birds are not the living descendants of dinosaurs, are they cousins of dinosaurs, derived from a more ancient common ancestor of birds and dinosaurs alike?

Darwin himself was bothered by the many obvious anatomical distinctions between birds and the other animals whose young develop as embryos in sacs, whether enclosed in shells or not—a common trait of the evolutionary group of amniotes, to which birds, mammals, and reptiles belong.

The publication of Darwin's *Origin of Species* in 1859 launched a battle between his supporters and antievolutionists, who pointed to the lack of intermediate forms—"missing links"—in the fossil record connecting major classes of animals, as evidence that evolution had not occurred. Just two years later, such a missing link was found by quarrymen in Solnhofen, Germany. The fine-grained, 140-million-year-old limestone of Solnhofen separates like pages of a book. Treasured by printmakers for its unrivaled image-retaining properties, this limestone has been mined for centuries. In 1861, workmen pried apart two sheets to reveal the most primitive birdlike fossil yet known, *Archaeopteryx lithographica* ("ancient wing from lithographic limestone"). Indeed, this flattened but spectacularly well-preserved fossil looked winged. Even the feathers on the crow-sized animal showed in the stone.

But aside from feathers, not much of *Archaeopteryx* is especially birdlike. It had teeth, claws, and a bony tail, all primitive characteristics that have been lost on modern birds. These peculiarities on a winged creature suggested that *Archaeopteryx* was an evolutionary intermediate between reptiles and birds.

The question of bird ancestry was, for most of this century, considered resolved by a Danish amateur ornithologist and anatomist, Gerhard Heilmann. In 1926, Heilmann published his book, *The Origin of Birds* (a 1916 Norwegian version had gone nearly unread). Heilmann observed that slim, early, predatory dinosaurs might, at least superficially, appear to be the most likely candidates to be bird ancestors.

However, Heilmann noted that these dinosaurs lacked that wishbonelike structure representing a pair of clavicles that anatomists call a furcula. Since earlier reptiles had these wishbones, as did all later birds, Heilmann decided birds could not be descended from dinosaurs. The birdlike characteristics of dinosaurs must, he reasoned, be a case of convergence, or evolutionary emergence of similar but unrelated anatomical features among different organisms. (Recall Stephen Jay Gould's example of the panda, which, in a series of adaptations enhancing its ability to grasp the bamboo plant it feeds on, has developed a thumblike digit that is not a true thumb but an extension of its wrist bone.)

Convergence, not dinosaur ancestry, might explain why birds, their wishbones or furcula excepted, so closely resembled some later dinosaurs. Instead of descending from dinosaurs, Heilmann concluded, birds developed from an earlier reptilian ancestor nearly 250 million years ago among the thecodonts. Thecodont is a common, if vague term, for the earliest members of the superorder to which birds, crocodiles, dinosaurs, and those reptilian fliers of dinosaur days, the pterosaurs, belong.

A hole in Heilmann's argument opened in 1936 when clavicle bones were discovered on an Arizona dinosaur nearly 200 million years old. Furculae have since been found on some other small predatory dinosaurs, chiefly from Mongolia.

Despite these discoveries, Heilmann's rejection of dinosaurs as bird ancestors, primarily based on the telltale wishbone, stood largely unchallenged during sleepy years for dinosaur paleontology, from the 1930s through the 1960s.

Birds From Dinosaurs

In the 1970s, however, a surge of new attention invigorated the inquiry into bird ancestry. Alick Walker, a British paleontologist, suggested in 1972 that birds arose from an ancestor of crocodiles. Birds do indeed have several anatomical features in common with crocodiles, their closest living relatives.

For a decade after, several paleontologists came to Walker's defense. But even more opposed him, and he ultimately rejected his own hypothesis in favor of what has become the consensus, that carnivorous dinosaurs, specifically, the small, fleet-footed, bipedal theropods, gave rise to the first birds.

A new dinosaur discovery forced a reevaluation of the relationship of dinosaurs to birds. In central Montana in 1964, John Ostrom of Yale uncovered *Deinonychus*—a "terrible-clawed," agile, and man-sized predator—and revived the dinosaur-bird connection theory. This dinosaur leapt and kicked and so had to have had a metabolism more like the active, warm-blooded birds than the sluggish, cold-blooded reptiles with which dinosaurs were then associated. In 1973, Ostrom argued that *Deinonychus* and other dinosaurs of its kind shared a recent common ancestor with birds among more primitive dinosaurs.

If birds were not dinosaurs, they were more closely related to theropod dinosaurs than to any other creatures.

Ostrom's brilliant undergraduate student Robert Bakker and British colleague Peter Galton went further, suggesting that birds are dinosaurs, both part of the superorder they named Dinosauria. "Dinosaurs are alive today," wrote Bakker in a provocative paper. "We call them birds."

Paleontologists were not inclined to accept the radical new Bakker-Galton unification of birds and dinosaurs (and pterosaurs, for that matter) into a single class, as it lacked a solid cladistic or other analytical foundation. But nearly all paleontologists were prepared to acknowledge the strengths of Ostrom's case for bird origins among dinosaurs. Ostrom drew particular and telling analogies between the anatomical features of the meat-eating theropod dinosaurs and those of *Archaeopteryx*. Both had the same structure in the forelimbs, hind limbs, and feet. Recent discoveries have illuminated many anatomical features shared by the later meat-eating dinosaurs and birds, most notably the braincase of *Troödon*. Analysis revealed not only that this dinosaur was the smartest creature known in dinosaur times but that *Troödon* had features, including a passageway across the roof of the skull connecting the inner ears, previously known only in birds and crocodiles.

In the last decade, cladistics has also come into play in the dinosaur-bird debate. Cladograms are very handy when applied to data as scattered, confusing, and variously interpreted as that from the fossils of dinosaurs. So, too, with fossil birds. Until recently, only two other birds, *Hesperornis* and *Ichthyornis,* were well-known from dinosaur days. Both of these date from the Late Cretaceous period, less than 100 million years ago. In the past few years two birds nearly as old as *Archaeopteryx* have been found, one in Spain and the second in China. The Chinese bird, nearly 135 million years old, is only eight inches long and has clawed fingers like *Archaeopteryx*'s. This discovery helps to close the gap not only between *Archaeopteryx* and modern birds, but *Archaeopteryx* and dinosaurs, for it "has many dinosaurian features including its stiff, short tail," according to Paul Sereno, who wrote on the newfound fossil in 1990.

Cladistic studies in 1982 by another former Yale University student,

Kevin Padian of the University of California/Berkeley, fortified the argument for dinosaurs as birds' ancestors. A more detailed cladistic analysis of dinosaurs by Jacques Gauthier, himself a former student of Padian's, discovered more than eighty characteristics that unite birds and small, predatory dinosaurs, from hollow bones to the anatomy of the hands and feet. Birds and dinosaurs were members of the same evolutionary group, by this cladistic analysis, just as Bakker and Galton had argued from a few features.

In the past decade most other paleontologists, but not all, have come to accept this line of reasoning. As critics of cladistics point out, a cladogram is only as good as the information that goes into it. Characteristics that seem advanced may not be advanced. A parallel response to the environment may lead to the convergent evolution of similar structures in quite unrelated animals; for example, bats and pterosaurs have wings. And cladistics does not directly consider the role of time in evolution. Small predatory dinosaurs may have many birdlike features, but the most birdlike of dinosaurs appeared some 70 million years after *Archaeopteryx*. This inconsistency suggests that the dinosaurs could not have arrived at birdlike features earlier than the appearance of the first birds.

With such limitations of cladistics in mind, in 1980, Queens College paleontologist Max Hecht and his protégé Sam Tarsitano argued that an as-yet-unknown advanced reptile was the ancestor of birds. They rejected the cladistic arguments. "Garbage in, garbage out," says Tarsitano of the cladistic analysis, asserting that many supposed similarities of dinosaurs and birds are more apparent than real. For example, the modern bird hand consists of the middle three digits while theropod dinosaur hands (and *Archaeopteryx*'s) are made of the first three fingers.

The same objections, and many others, are now offered principally by one vocal and persistent critic of a dinosaurian ancestry for birds, Larry Martin, a bird paleontologist at the University of Kansas. Martin is widely acknowledged as a world expert on fossil birds.

For rejecting the popular thesis that dinosaurs are ancestors of birds, Martin is known as "the Antichrist of the bird-origins business," as he jokes. Martin puts his view of the bird-dinosaur relationship bluntly: "I don't think birds derived from dinosaurs. The dino argument doesn't stand up. You stick a finger into it, it goes right through."

Martin has been working for years on a voluminous paper rejecting each of Gauthier's scores of alleged shared characters between dinosaurs and birds. "The people who argue that birds are related to dinosaurs use features that aren't generally found in all dinosaurs. They're found as isolated occurrences in certain specific dinosaurs, and most of these dinosaurs are very late in the geological record. So these paleontologists have a tendency to try to derive birds very, very late," says Martin.

Martin was once part of that paleontological mainstream believing in the dinosaur ancestry of birds. But thanks to a graduate student's observation, Martin began questioning the dinosaur-bird relationship a decade ago. The student found that teeth in *Ichthyornis* were identical to crocodile teeth in structure and in the way they were implanted and replaced. "I've been looking for a dinosaur tooth like it ever since, and haven't succeeded," says Martin.

So, on the basis of teeth, Martin suggests that birds, like crocodiles, branched off from the line that would lead to dinosaurs. Birds may have gone their own evolutionary way even before the last thecodonts evolved, perhaps more than 250 million years ago.

All parties in the bird-origin dispute consider their debate at an impasse. The weight of scientific opinion, if not evidence, all would concede, now lies with the view of birds as derived from dinosaurs. That weight can press hard on dissenters. Says Martin, "Walker recanted his theory on crocodiles as birds' ancestors. I think opposing the popular viewpoint gave him a heart attack and a nervous breakdown. But I'm too pugnacious to care."

A Newer Older Bird

Both sides in the bird-origins debate are handicapped by a woeful shortage of fossil data, particularly of early birds. Recent finds have revealed a bit more of early birds and birdlike dinosaurs, but none older than *Archaeopteryx.*

None, that is, except perhaps a few dozen shards of white bone lately found in the arid scrub plains of the Texas panhandle by paleontologist Sankar Chatterjee. Reassembled and aligned, these fossils comprise most of the skeleton of a distinctive creature. If Chatterjee has interpreted it properly, it may be the most significant paleontological find of the 1980s.

To the ebullient and confident Chatterjee the bits of hollow white bone he uncovered in 1985 from red clay and sandstone hillocks are, without question, the remains of the earliest known bird in prehistory, as much as 225 million years old. Chatterjee's *Protoavis* is 85 million years older than *Archaeopteryx,* a yawning gap in time wider than that which separates us from *Tyrannosaurus rex.*

Six years after making his discovery, Chatterjee published his lengthy and long-awaited scientific analysis of *Protoavis* in 1991. Yet his assertions still have few supporters among the few of his North American colleagues who've seen *Protoavis* and the many who've only heard and read of it. (Support for *Protoavis* is stronger among European paleontologists.) It is possible that Chatterjee is correct in his claims for *Protoavis,* and that the origin of birds from *Archaeopteryx,* or from any dinosaurs at all, is a fallacy.

Acceptance of Chatterjee's interpretation of *Protoavis* would change *Archaeopteryx*'s status from the earliest known bird to a rather late and primitive experiment in dinosaur-to-bird evolution, nothing more than a curious evolutionary dead end.

Moreover, Chatterjee's *Protoavis,* if accepted, pushes the date of the origin of birds back by half, making the first birds nearly contemporaneous with the first dinosaurs. Under such a scenario, it is all but impossible for dinosaurs to have been ancestral to birds. And while it would seem improbable at best that a bird could be found in strata 225 million years old, to be followed by nothing remotely birdlike until *Archaeopteryx* 85 million years later, the fossil record is full of holes. The evidence of ancient birds found to date, even after *Archaeopteryx,* is minimal.

Chatterjee has avoided speculating on the evolutionary significance of *Protoavis.* He doesn't see *Protoavis* as a "missing link" in bird evolution. There is, in fact, no such thing as a "missing link," a single

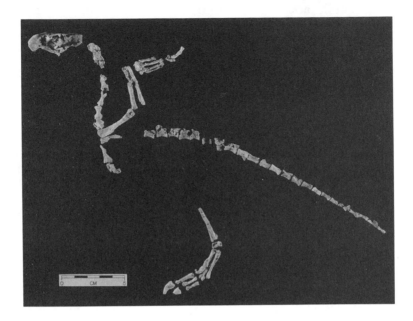

The restored bones of Protoavis, Sankar Chatterjee's controversial candidate for, by far, the world's oldest bird. [Photo courtesy Texas Tech University]

transitional ancestor in the evolution of any living creature. The possibility always exists that we will discover one or several fossils that appear ancestral to those creatures we have ascribed to a particular group.

But if Chatterjee is right about *Protoavis,* birds are not dinosaurs nor their descendants. Rather, birds evolved independently of dinosaurs from a primitive reptilion ancestor of birds and dinosaurs. The evolutionary pathways drawn up by Gauthier and Padian and other cladistically minded paleontologists would be invalidated.

However, if Chatterjee has erred in interpreting his find—and several of the most distinguished dinosaur paleontologists believe he is thoroughly mistaken—then he's confused one of the most important turns in evolutionary history.

The details of Chatterjee's find, of his own spectacular career, and of the slim fossil record for all of life, birds in particular, both question and confirm *Protoavis*'s significance. The questions of bird origins among dinosaurs, and of *Protoavis*'s role in the evolution of birds, may now be even more difficult to answer, but they are still well worth considering.

Cry Wolf?

The discovery of *Protoavis* is not the first occasion on which, as Chatterjee's admirers would point out, he's made spectacular discoveries. Nor the first occasion, as his detractors would point out, when he's derived a dramatic, unorthodox, and perhaps wrongheaded conclusion.

Chatterjee has uncovered many odd and unfamiliar fossils—from primitive sauropods in India to early prosauropod dinosaurs in China to dinosaur-aged fish and plesiosaurs in Antarctica. Chatterjee smiles and passes off such success as serendipity. "Mind you, it's just luck for me," he says modestly. For other diggers, though, he'll posit a mystical talent, such as the unschooled Indian villager he worked with who could "hear" fossils below the ground. Or the legendary and still-active paleontologist James Kitching of South Africa, who spotted productive fossil beds from a moving train in the African desert and from an airplane window over Antarctica. Or the late Barnum Brown, longtime "dinosaur man" of the American Museum of Natural History. It could be said of Chatterjee, as American Museum of Natural History president Henry Fairfield Osborn said of Brown, "he must be able to smell fossils."

Scientific training, however, not a sixth sense, gave Chatterjee the opportunity to look for fossils. Born in Calcutta, in 1943, to an educated family (his father was a chemist, his brothers, engineers), Chatterjee trained as a geologist at the most prestigious scientific center in the nation, the Indian Statistical Institute.

From the start, Chatterjee had a knack for finding fossils. Among his early finds, in central India, were complete skeletons of two ancient crocodilelike reptiles, and curled within their stomach cavities, the remains of their lunch—two smaller fossil reptiles. "Nobody finds anything like that," says Smithsonian Institution paleontologist Nicholas Hotton III. "Just one whole skeleton is very hard to come by. Even when we know where to look, one in five years is good going. . . . If finding fossils is luck, Sankar's the luckiest guy in the world."

But paleontology offers few permanent jobs, even to lucky guys. After several fellowships in far-flung locales, Chatterjee landed a job in 1979 as a geology professor at Texas Tech University in Lubbock.

As impoverished a profession as it is, paleontology does have its well-funded, highly regarded centers of learning. Berkeley is one of them. Lubbock is not. The university, dominated by its football stadium, squats in the midst of a drab city of some 200,000 splattered across the barren West Texas plains.

Sankar Chatterjee's office looks out on none of this alien world. His is a windowless suite in the basement of the museum, or as it is plainly titled, The Museum. By any name the place is a large and lifeless sandy brick rectangle, which thanks to an adjoining silolike planetarium, looks more like a nuclear power plant than a scientific institution. The museum receives no state funds, only donations from the public.

All of this means that Chatterjee and his two graduate students haven't got much more than a vehicle to further their research—no funds for far-flung research, nor for trained preparators to do the laborious job of separating fossils from the rock matrix in which they are found. Like paleontologists everywhere outside of the few well-endowed institutions, he must look almost entirely to outside sources, specifically the National Science Foundation, for funding.

Chatterjee has been successful in obtaining research support. On the door to Chatterjee's office, a small sign reads ANTARCTIC RESEARCH CENTER. The university has a long tradition of polar research. Now, "I'm it," says Chatterjee, laughing, "when I can get grant support to go." He's done so three times, journeying to Seymour Island, and nearly sinking in 1987 in an iceberg-torn ship.

Much closer to campus, Chatterjee has continued his remarkable record of locating significant fossils. His ongoing digs are centered in

Sankar Chatterjee in the badlands outside Post, in western Texas, where he discovered Protoavis and many other intriguing fossils. [Photo by author]

the hill country fifty miles southeast of Lubbock. On a ranch outside of Post, a cowboy town of 4,000 (and a failed utopian grain-growing community of cereal magnate C. W. Post), Chatterjee began digging in a formation previously explored by several paleontologists. Those prospectors had located fossils some 235 million years old, from predinosaurian Early Triassic days, in these red beds of the Dockum Formation.

In his forays, Chatterjee hit upon a strata of rock some 10 million years younger. The exact age of the sediments is difficult to determine, but the distinctive nature of the fossils Chatterjee encountered makes the relative dating convincing. What Chatterjee came upon in these younger rocks was a whole new population of animals previously unknown from the Dockum. Chatterjee found abundant evidence of a diversified animal life. He uncovered remains of pterosaurs and of the big, agile, crocodilelike reptiles that were the dominant hunters of the era. He found the tiny ankle, jaw, teeth, and vertebrae of a bird-hipped dinosaur no taller than a ten-year-old child. These bones, according to Chatterjee, belong to the earliest known ancestor of the duckbilled, horned, and armored dinosaurs, a dinosaur he named *Technosaurus* in honor of Texas Tech.

Significant as these discoveries are, they have not attracted universal professional acclaim for Chatterjee. His association with a small paleontological institution, his paucity of graduate students and former graduate students, all leave him outside the profession's mainstream. Moreover, Chatterjee's penchant for grand discoveries, and even grander speculations on their significance, have distanced him further from his colleagues.

Chatterjee has attempted several significant theoretical contributions to an understanding of the prehistoric world, including a controversial reassessment (with paleontologist Nicholas Hotton III) of the ancient position of India. India was long thought to have been an island for most of dinosaur time before slamming into Asia, creating the Himalayas (a mountain-building process that continues to this day at a rate of six inches a year). Analyzing dinosaur remains and other fossils, Chatterjee and Hotton concluded the animals were so like their contemporaries in the northern continents that India must have been linked to Africa and then Asia for most of dinosaurian times. Geologists found Chatterjee and Hotton's evidence far from sufficiently persuasive to overwhelm the impressive evidence of sea-floor spreading that points to India's position as an island continent in the age of the dinosaurs.

Chatterjee did help define, in a 1982 analysis, the evolutionary changes that created that central characteristic distinguishing dinosaurs from earlier reptiles, the simple and rigid dinosaurian ankle. That rather technical anatomical analysis has been disputed by some colleagues who assert it is based on faulty interpretation of incomplete fossils.

And in India, in 1988, Chatterjee and Indian colleagues explored dinosaur-bearing sediments in Jabalpur in central India. There the researchers found what Chatterjee describes as "beautifully preserved partial dinosaur skulls." But to Chatterjee and his co-workers what made the find of great import was another discovery, of shocked quartz—an aberrant form of the crystal commonly thought to result only from the singular force of an extraterrestrial object striking the earth. Chatterjee thinks he may have found the "smoking gun," the place where the asteroid struck leading to the denouement of the dinosaurs. Again, few of his colleagues appear to have accepted Chatterjee's interpretation, and the search for an explanation to account for the dinosaurs' death has gone on elsewhere (see Chapter 11).

Chatterjee's skills in interpreting his own remarkable fossil finds have also been questioned. Chatterjee identified one fossil as the jaw of a mammal-like reptile. Other paleontologists subsequently disputed Chatterjee, identifying the fossil as a fish scale. In a 1991 paper, Paul Sereno found that *Technosaurus* not only came from more than one individual animal, but from more than one kind of dinosaur and that Chatterjee had juxtaposed and confused several bones.

Chatterjee had compared another of his Texas discoveries, the crocodilelike *Postosuchus,* 225 million years old, to dinosaurian carnivores such as *Tyrannosaurus rex,* which lived 70 million years ago. Chatterjee concluded that the tyrannosaurs evolved from such early reptiles. To his colleagues, comparisons made over such a wide gulf of time were meaningless, and as one paleontologist said dismissively, "wild."

So it was with even more than characteristic professional skepticism that vertebrate paleontologists received the astounding news in 1985 that Sankar Chatterjee had come upon what he believed to be the first bird, a bird that predated *Archaeopteryx* by 85 million years.

In the spring of 1985, Chatterjee was picking over crumbling, pebbled sandstone at a Dockum mound he'd already shoveled away when he spotted a scattering of delicate white bones, from one or several unfamiliar little creatures. The white shards of leg bone that he teased from the soil with a dental pick and brush looked to him at first "like a little dinosaur I'd found in India. It had a dinosaurlike ankle joint," a simple but strong hinge of the sort found in pterosaurs, dinosaurs, and birds.

But as he dug further, and in his laboratory, carefully cleaned away the bone from mudstone, Chatterjee identified pieces of skull, pelvis, and vertebrae of two individuals of the same unknown species, each smaller than a crow. The delicate proportions of an ankle bone, less than one-quarter inch in diameter, suggested the smaller of the two might be a juvenile animal. But baby bones were never fused, as these were. These were adults, but adults of what animal?

The ankle structure was a rigid one, unlike those of crocodiles, the dominant animals of the time. The holes within the pelvises and skulls were peculiar to birds. The creature was so much a bird, Chatterjee thought he'd found a wishbone, or half of it, in a spatula-headed fragment of bone. Wishbones are found in birds but not in early crocodilelike reptiles. Wishbones have been found on several of the later predatory dinosaurs, but none are known from a dinosaur within 100 million years time of the Dockum Formation.

The wishbone was Chatterjee's first strong clue that what he had was a bird. Chatterjee also thought he saw faint pimply bumps on frag-ments of forearm bone of his birdish find: quill nodes for the attach-ment of feathers. He also detected a hollowing of the ends of the vertebrae, a shape usually found only on bird backbones.

A conservative scientist might have concluded that he had found a dinosaur, albeit an unfamiliar small one with several curious features. Certainly, the shape of the ankle joint suggested this was far more dinosaur than crocodile. But the more Chatterjee cleaned and studied the creature under the microscope, the more he thought it looked like a bird. Perhaps most telling of all was a hollow, flexible piece of quadrate, a jaw bone so constructed only in birds.

The skull of the fossil was, to Chatterjee, unquestionably birdlike. In the birds' skulls, as Chatterjee points out, many features have been dramatically modified from their reptilian ancestors'. "Birds are able to raise their upper jaw to increase the gape, but no dinosaurs could do this. To achieve this upper-jaw mobility, the temporal region of the bird skull has been modified considerably." In Chatterjee's find, the bony struts of dinosaur jaws had disappeared, producing a birdlike and flexible jaw. The braincase was enlarged compared to that of previously known early dinosaurs, indicating to Chatterjee that this creature had specialized nerve centers for flight.

Chatterjee compared these and other skull features with those known from *Archaeopteryx,* and from *Troödon,* the most birdlike of the late theropods, a dinosaur from 75 million years ago. He found twenty-three distinctive avian features in his creature's skull that are not found in *Troödon* or any theropod, or crocodilian for that matter. On the basis of such detailed study Chatterjee concluded he had "a true bird."

Surprisingly, to him, most of these birdish features in the skull of his find were absent in *Archaeopteryx. Archaeopteryx,* he reasoned, was like *Troödon,* an example of convergence—evolution shaping disparate organisms toward a common end.

In the case of *Troödon* and other Late Cretaceous dinosaur predators, the development of keen senses, a relatively large brain, and nimble hands for hunting also produced birdlike features in the skull and limbs. *Archaeopteryx,* with its bird feathers but long, reptilian tail, constituted, to Chatterjee's mind, another evolutionary move toward flight, as bats did among mammals, or pterosaurs among reptiles. This particular experiment may well have ended with *Archaeopteryx.* Perhaps the *Archaeopteryx* experiment was even an offshoot of the evolutionary innovation that produced Chatterjee's birdish fossil 85 million years earlier. *Archaeopteryx* may have evolved near that time, meaning that the *Archaeopteryx* fossils we know may have come from a persistent primitive creature that lived on until 145 million years ago. Regardless of *Archaeopteryx*'s age, the bush of evolution had already branched off toward modern birds before *Archaeopteryx,* Chatterjee reasoned.

He is convinced that the branching passed through, if it did not originate in, the creature he found. He proudly, if gingerly, pulls its skeleton out from a metal file cabinet at the request of interested visitors. Here, arrayed in a form strikingly like that of a small modern bird, are the white bones of Chatterjee's find, with more links than gaps in the skeleton.

Informally, Chatterjee dubbed the creature "Protoavis," or "first bird." Formal naming had to await peer recognition of his published scientific description and analysis of the find, six years after his discovery.

It took that long for Chatterjee to clean, arrange, and analyze these minute and delicate bones. But long before he'd removed most of the bones from their rocky matrix, Chatterjee began showing many of the fossils to colleagues. First to view them was his friend and former mentor Hotton. Hotton agreed "it seemed to have some birdlike qualities, but it was too weird for me to know what it is."

Next to see *Protoavis* was John Ostrom, the Yale dinosaur paleontologist who'd discovered the agile dinosaur predator *Deinonychus* and identified the sixth *Archaeopteryx* specimen. Ostrom was called in by the National Geographic Society as a cautious senior scientist with long-standing interest in bird origins—neither a young cladistic theoretician nor an old-style field man such as Chatterjee. Ostrom didn't see quill bumps, nor was he sure about the shape of the vertebrae. But the quadrate, the distinctive jaw bone, was to his mind strikingly birdlike, if incomplete.

The strongest endorsement of *Protoavis* as the earliest bird came from bird paleontologist Larry Martin. Martin invited Chatterjee to bring the *Protoavis* material to the University of Kansas for a week. Martin detected thirty-one characteristics in *Protoavis* that were similar to those of modern birds. Among these were the mobile quadrate, and a pelvis with a large indentation, perhaps intended to serve, as such pockets do in modern birds, to protect enlarged kidneys. From his study, Martin concurred that *Protoavis* was a true ancestor of birds.

Martin acknowledges that the discovery of a birdlike creature 225 million years ago in the Late Triassic buttresses his theory of bird origins among creatures that lived even earlier, before the advent of dinosaurs. "I would be quite happy to have someone find a bird or a close bird relative in the Late Triassic."

But Martin says his support for *Protoavis* isn't based on its neat fit with his bird evolution theories. "Sankar's 'Protoavis' has a whole suite of characters that are otherwise only found in birds, and are not demonstrated in any other groups of dinosaur, which I think unequivocally shows that '*Protoavis,*' whatever it is—and I'm not convinced that the thing had feathers and flew—is a closer relative to birds than anything else that's been presented. That's quite different from calling it a bird," Martin cautions.

Another more celebrated iconoclast among paleontologists, Robert Bakker, examined *Protoavis* and came away similarly impressed, if more puzzled than Martin. "It's one hundred percent primitive theropod, but one hundred percent primitive bird. The braincase is very birdlike, the air sacs very birdlike, the quadrate strut most impressive. It's really remarkably like a modern bird in the quadrate—a complex little gizmo with a pivot joint. I just saw the good *Troödon* material, and this is a lot more birdlike in the quadrate than *Troödon*. Is '*Protoavis*' a dinosaur or already a bird? I don't know."

But some of the handful of others who've been to Lubbock to see *Protoavis* are far less impressed. Cladist Jacques Gauthier, for one, was immediately disappointed. One might have predicted as much, since Chatterjee did not subject his find to the rigorous cladistic analysis in which Gauthier was trained at the University of California/Berkeley. According to cladistic analysis, birds evolved from dinosaurs, and in this respect *Protoavis* is completely out of place. Yet Gauthier's reasons for disputing Chatterjee's find were compelling and specific. "I'd seen Sankar's reconstructions—and by those drawings it was clearly a bird. . . . The specimen didn't look anything like the drawing." As for the "partial wishbone" Chatterjee had spotted on *Protoavis,* Gauthier saw it as a complete arch from the tail vertebra of a primitive reptile.

Gauthier didn't disparage Chatterjee's analysis, but he did say that "Sankar overinterprets poorly preserved parts, finding what he wants to see in them. '*Protoavis*' is a particularly egregious example of what's called 'the sympathetic eye.' " What exactly it was, Gauthier is uncertain. "If Sankar is right, it's a flighted dinosaur. Flight has arisen before birds [pterosaurs] and since [bats]. But it seems like an odd, small new species of dinosaur to me."

Chatterjee says that since Gauthier saw the skeleton, new bones have been discovered, and the skeleton is now 80 percent complete.

But if Gauthier hasn't seen the full evidence for *Protoavis,* neither have other paleontologists. And many of them are irritated by Chatterjee's popular announcements of his find. Chatterjee hasn't gone the standard collegial route, waiting until his research was done, making his fossils and findings available for close inspection before publication.

When Chatterjee pulls the case containing *Protoavis* from its cabinet, he rests it on his desk, directly atop one of the many quotes he's pasted around his office, among them a vow by another dedicated field man, Charles Darwin, "never to engage in controversy, for [it] rarely did any good and caused a miserable loss of time and temper."

What Darwin found in the Galapagos and elsewhere, however, was bound to cause a furor. And though he delayed publishing his views on natural selection, Darwin did eventually come forward.

Chatterjee, likewise, says, "I hate controversy." He maintains that he would have waited to publicize his discovery until his research was completed, but his longtime research sponsors, the National Geographic Society, thought an announcement was in order. Ostrom's tentative endorsement of the fossil's worth had satisfied the Geographic, and a Washington press conference was held in August 1986 to herald the find. Extensive press attention followed immediately, including John Noble Wilford's article in the *New York Times*—titled "Texas Fossil May Be Birds' Oldest Ancestor," and featuring an artist's rendering provided by Chatterjee of a reconstructed skeleton with superimposed wing, and an endorsement by the Smithsonian's Hotton: "It looks bird."

So, the differences between Chatterjee and the critics of *Protoavis* go far beyond the fossil itself. They are differences between modern cladists and an old-fashioned field man, between establishment and outsider, between an accepted theory and puzzling heresy.

Six years after he found *Protoavis,* Chatterjee says, "This is a complicated question and I wanted to get it right." First drafts of this *Protoavis* treatise went out to colleagues for review in the summer of 1989. That a published account did not appear until two years later is not an unusual delay in the slow-moving realm of academic publications. (Indeed, the journal that published Chatterjee's paper, *The Philosophi-*

cal Transactions of the Royal Society of London, is a well-respected scientific journal.)

Even Chatterjee's friend Hotton allows, "I think Sankar went about it wrong. He has a product to sell. He's got to present it in the strongest light. Maybe he just should have said, 'I've got an interesting small dinosaur here, maybe the smallest one of all, maybe ancestral to birds, and let others support me.' But that's not like Sankar, he's gone for broke."

The criticism of Chatterjee's radical but unpublished conclusions about *Protoavis* has taken on an unusually acid personal tone. University of California/Berkeley professor Kevin Padian boldly criticized Chatterjee's approach in print. Padian is a well-trained cladist himself. He is also a man of strongly voiced opinions. He included *Protoavis* in a paper on misrepresentations of paleontological science, concluding: "Few people have seen the material, though those that have are largely skeptical that it represents a single animal, and it is difficult to confirm his interpretation that any of the material comes from a bird." These were unusually bold charges about a colleague, and a work not yet published.

Larry Martin jumped to Chatterjee's defense, calling Padian's criticism a regrettable instance of "personalizing a debate and ignoring the evidence that conflicts with his position."

Personal and institutional biases aside, the central problem for Chatterjee to overcome is that *Protoavis* simply doesn't fit into the prevailing view of bird ancestry. "If all Sankar says he's got is true, he's done it," says eminent dinosaur paleontologist Peter Dodson of the University of Pennsylvania. "But all of this sounds too much—like he's got the remains of yesterday's chicken dinner."

At heart, the bias that may prevent Chatterjee's *Protoavis* from getting its due is neither personal nor institutional, but professional, as Larry Martin opines. "Most everyone else who's looked at 'Protoavis' works on fossil reptiles. They don't know avian anatomy very well. When you look at new material, stuff that you're not familiar with, you key in on the things you've seen before. If somebody shows you a map of the United States, your eyes will travel to the places that you've been. When

I looked at *Archaeopteryx,* I was impressed with how many avian features it had. When a dinosaur paleontologist looks at *Archaeopteryx,* he has a tendency to be impressed with what looks like dinosaurs. I think that's been part of the problem."

Protoavis may yet get its fair hearing, now that Chatterjee has presented his data and analysis, when others have followed with theirs. But there is no assurance in this science that the best explanation for bird origins will win out. There is far more prejudice and uncertainty in science than scientists care to admit. And the most elegant hypotheses in science are compelling only until a new set of facts is uncovered, facts that require a fresh explanation.

In October of 1990, Chatterjee came to the fiftieth anniversary meetings of the Society of Vertebrate Paleontology in Lawrence, Kansas, with another puzzling fossil he'd found in the 225-million-year-old red clay of West Texas. This was a nearly complete skull, six inches long, with a bladelike beak, entirely toothless. The creature bore no resemblance to the dominant reptiles of its day. The delicacy of its skull was reminiscent of birds. And its size and toothlessness put several paleontologists in mind of ostrich-mimic dinosaurs, a group thought not to have evolved for another 100 million years. The specimen was so little plastered that the accuracy of its appearance was unquestioned. But its identity was. Chatterjee, not cowed by bitter experience, speculated in a presentation at the 1991 SVP meetings that this mysterious skull was the first ostrich-mimic dinosaur. Colleagues were skeptical. But the discovery served as a reminder, nonetheless, that many evolutionary experiments were going on at the dawn of the dinosaur age, experiments that might include a first bird such as *Protoavis.*

Chatterjee has yet to establish *Protoavis* as a central fossil in bird evolution. But *Protoavis* has already provided enough fossil data to point out, to those who care to admit it, that all the evidence of ancient birds adds up to very little, and that the case for a dinosaurian origin for birds is still shaky, as are all other scenarios for how birds evolved. In time, mounting data and subsiding partisanship will permit a clearer view of *Protoavis.*

Chatterjee takes just such a long view. So, despite his aversion to the controversy he has created, Chatterjee would do it all again. "I am sorry

Protoavis caused so much bitterness," he concludes, "and sorry it has taken me away from the field for so long. But if *Archaeopteryx* can take more than a hundred years to interpret, one should allow at least ten years to understand *Protoavis*. It is a very complex evolutionary and taxonomic problem, which I think I will be able to resolve, if I get sufficient time."

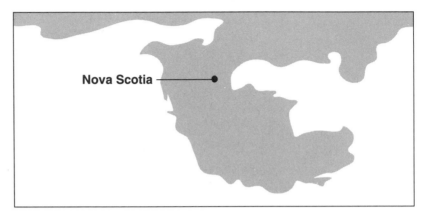

Pangaea, 215–200m. yrs. ago

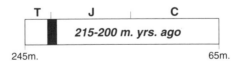

245m.   65m.

AGE OF THE DINOSAURS

Triassic Jurassic Cretaceous

First animals   First land animals   First human

	T	J	C	
590m.	350m.	245m. 208m.	145m.	65m. Today

The Dinosaur Takeover

The first dinosaurs, *Herrerasaurus* and perhaps *Protoavis* among them, were only subsidiary players in an equable world-continent more than 200 million years old, a stage dominated by giant reptiles. So when, and how, did the dinosaur assume top billing in the animal world? Paleontologists have long theorized about the timing and circumstances of the dinosaur takeover. Only lately has a fresh batch of evidence suggested new answers to these central questions of dinosaur evolution.

The proliferation of dinosaurs, and the disappearance of their competitors, was, until quite recently, thought to have transpired slowly, taking until the end of the Triassic period, around 200 million years ago. The lizard-hipped dinosaurs, those with a forward- and down-pointing pubis bone, expanded in numbers and kinds from relatively small two-legged carnivores to include bulkier two-legged herbivores. At about the same time, an anatomical innovation produced other small plant-eaters of the bird-hipped variety, those with a backward-pointing pubis. These two dinosaur lines produced varied and successful forms of dinosaurs for more than 100 million more years before their extinction.

In the Jurassic period (to 145 million years ago) four-legged, lizard-hipped dinosaurs attained record sizes for all land animals, more than one hundred feet long for some of the plant-eating giants. In the subsequent Cretaceous period (to 65 million years ago), dinosaurs evolved into other spectacular forms, among them duckbilled bird-hipped dinosaurs and forty-foot-long lizard-hipped tyrannosaurs, the largest predators ever.

The long-standing view of dinosaur ascendancy is that they some-how won out over their contemporaries by long evolution that might have involved large-scale "competition" between old and new animals. The dinosaurs triumphed, it was thought, because of their naked skin, which was better suited to dissipate heat in the hot, dry climates of their day. Or they survived preferentially because of their upright stance, which permitted them to eat off higher branches, and therefore with less competition. Or perhaps via their metabolism, which enabled them to run fast longer than semierect or sprawling cold-blooded competi-tors.

Now it appears that the rise of the dinosaurs occurred with relative swiftness, that is, within several million years. And the emergence of the dinosaurs may have been, according to one recent theory, no more than a chance development.

One or even two of the many extinction events in earth's history may have been responsible for the rise of the dinosaurs very early in their history, just as a sudden mass extinction may have finished them off nearly 150 million years later.

A widespread extinction marks the Triassic-Jurassic boundary, 200 million years ago. By many accounts, that boundary occurs at one of the most significant and puzzling of some thirteen or more major extinction events in the history of life on earth. It was a time of drastic change in the environment of earth's lands—the supercontinent split in two, a massive faulting accompanied by lowered seas, more extreme climates, and rampant volcanic eruptions. After this epic juncture, many early amphibians and reptiles disappear from the fossil record. Among them were the dominant land animals—many mammal-like reptiles, and many other reptilian families. What endured was a dino-saur-dominated world that also included turtles, frogs, salamanders, lizards, crocodiles, and mammals.

That boundary extinction, according to long-prevailing wisdom, marked the end of a long, slow decline for the dinosaurs' large competitors. This slide into extinction was induced by a protracted contest the dinosaurs had won, either by superior locomotory skills or by better adaptation to the dry, hot conditions that prevailed at the time.

But a young British paleontologist, Michael Benton, has pointed out flaws in both these arguments. The "mammal-like reptiles" already had a variable gait, so why would they be so devastated by the emergence of dinosaurs with their erect posture? As for the climatic-change scenario, Benton maintains that it is based on scanty, flawed arguments that the red color of Triassic rock was a sign of hot, dry conditions. Rocks from many environments turn red. We do not know, Benton asserted, what the climate was like 200 million years ago.

Benton went further, rejecting the entire competition explanation for dinosaur dominance. It was a bold assertion for a paleontologist who was only twenty-seven at the time. Benton reasoned that dinosaur success was a result of their good fortune in the face of a sudden mass extinction. This great extinction was far more widespread than the one at the Triassic-Jurassic boundary, 200 million years ago, and struck fully 20 million years earlier. From skulls and complete skeletons of diverse animals collected around the world, there seems to be no evidence whatsoever for dinosaur competition with other Triassic animals, says Benton. Rather, many families of early reptiles disappeared at the same time, 225 million years ago, before the diversification of dinosaurs. According to Benton, dinosaurs simply had the good fortune to survive, and to fill most of the niches vacated by these vanishing animals.

Many paleontologists and geologists have questioned Benton's "lucky break" analysis of this first mass extinction in dinosaur times. They note that this extinction was not synchronous among land and marine animals and not as devastating to large land animals as the catastrophe some 20 million years later. But Benton has compiled fossil data to support his case, now awaiting publication. Other paleontologists await these data on the timing of the changeover to a dinosaur-dominated world before casting their lots for or against Benton's theory.

Until lately, the many gaping holes in our knowledge of dinosaurs and their environment included not only the early years of dinosaur evolution but the portion of the Jurassic period during and after the time dinosaurs achieved dominance. Which animals survived the great extinction of 200 million years ago—the fauna of the earliest Jurassic—was until the last decade a complete mystery. "To put it succinctly," paleontologist Kevin Padian wrote, "there were simply no known Early Jurassic terrestrial vertebrates, except for the odd dinosaur or two washed into some marine sediments in England or Germany." From shortly after the time of the first dinosaurs, 225 million years ago, until the dinosaur-rich fossil beds of the Late Jurassic, 75 million years later, the evolutionary saga of the large land animals was unknown.

Now, however, the animals of 200 million years ago and their environment, at least in one part of the world, are sufficiently well known that dinosaur evolutionary theories as all-embracing as Benton's can be put to the test.

And now another paleontologist has a different "lucky break" theory to account for dinosaur ascendancy. Perhaps dinosaurs won out 200 million years ago because they survived a catastrophic impact of an object from space.

Paul Olsen is convinced of the veracity of that hypothesis. And in eastern Canada, Olsen believes he's found the proof. The prevailing climate 225 million years ago is unknown, but Olsen has found evidence of a changing environment 25 million years later. He has done this in stunning detail by bringing to bear a host of other disciplines including astronomy, chemistry, and dinosaur studies. The incorporation of these and other sciences into dinosaur paleontology is central to its current renaissance.

Nova Scotia today lies exactly halfway from equator to pole. Coastal Nova Scotia is home to many summer resorts, but Parrsboro, along the peninsula's west coast, is something else, a sleepy, plain town with several Victorian-era buildings and 1,800 hardy souls. For a disparate but congenial band of paleontologists, fall is a fine time and this is a fine place to look for fossils from early dinosaur days.

Most of the thousands of fossils these scientists have discovered along the cliff faces of Parrsboro represent small creatures or have been fractured to tiny and almost unrecognizable bits. Few are identified, even prepared. If put together, none would form more than a small part of a skeleton.

What makes these fossils most remarkable is that here, as virtually nowhere else on earth, it is possible not only to find unusual dinosaurs but to locate their place in time with impressive exactness. The rocks, and so the fossils within them, at Wassons Bluff outside Parrsboro, date to 200 million years ago, the Triassic-Jurassic boundary, when a new continent and a new world-ruling group of animals emerged. And nearby, also, is the site of an event that may yet prove that the reign of the dinosaurs, not their demise, was caused by an extraterrestrial impact.

It's a short drive from Parrsboro along a scenic cliff-skirting road to a small turnout and a dirt trail down to the beach at Wassons Bluff. To the east, stretching to the horizon, lies a ribbon of banded-gray cliffs of Early Jurassic and Triassic age, topped, as are those to the west, with evergreens and in autumn, flame-leafed birches. More than one hundred feet high, the cliffs wind behind the beach. Cones of gray-black basalt, extruded by ancient volcanic eruptions, rise toward the tree line, interspersed with mudstone, and broken by long, jagged diagonal fault lines, along which orange sandstone is packed. Nearby, broader bands of sandstone stretch from coast to cliff-top between basalt blocks.

At six sites in the cliffs along a mile of wild beachfront—some basalt, some pockets of sandstone, some entirely sandstone—Olsen and his crew of American paleontologists have been digging since 1985. The topography is complex, and in the long shadows of the afternoons the promontories seem to change shape as well as color. But the site has been notably productive, and the paleontologists have found a hundred thousand pieces of bone.

These include the limbs, vertebrae, and teeth of a chicken-size plant-eating dinosaur, and the hind end of a less than four-foot-long prosauropod, a two-legged herbivorous dinosaur common around the world in the early Jurassic. Isolated teeth were discovered from a small predator no more than six feet long. The slopes are flecked with white dots of fossil bone, succinctly compared to bird droppings by one

Nova Scotian dinosaurs [John Sibbick]

frustrated excavator. Only by painstakingly chipping away at the cliff-side can the paleontologists determine if a fleck represents the exposed edge of something larger and identifiable.

The orange zones in the cliffs preserve dunelike environments, cut by lakes and rivers. Between fissures in the basalt itself are the densest accumulations of animals. Slowly, 200 million years ago, silting material built up there, providing niches into which animal bones settled. In the basalt itself many more bones were found, badly broken-up.

Dinosaur prospecting around Parrsboro began when Eldon George, an amiable and long-winded local fossil enthusiast and vendor in his late fifties, found a set of dinosaur trackways smaller than any previously known. George thought the tiny tracks, the size of a penny, belonged to a crocodile, but American paleontologist Paul Olsen pegged them as dinosaur fossils.

Olsen is a talented amateur guitarist, carpenter, and professional scientist. He is trim and blond, and his blue eyes are framed by thick wire-rimmed spectacles. Olsen is somewhat of a mystery man to his colleagues, so wrapped up in research he rarely answers letters and phone calls. In his work, Olsen is at once a Renaissance man and a specialist's specialist—expert in geophysics, lake geochemistry, taxonomy, fossil footprints and morphology. But he employs all these disciplines to better understand one formation and one place in time: the Newark Supergroup, rock made from what lay along eastern North America some 200 million years ago.

After recognizing the dinosaur fossil footprints of Parrsboro in 1975, Olsen returned to scout outcrops with Don Baird, a Princeton paleontologist, and his preparator, a young Montanan by the name of Jack Horner. Horner and Baird were looking for fossils from predinosaurian times. Olsen went his own way, investigating a section of cliff face he'd never seen before. In cold and pouring rain he walked twenty miles from Five Islands to Parrsboro with a plastic sheet over his head. At Wassons Bluff he saw the badly fragmented remains of a dinosaur weathering on the beach. But never having seen a dinosaur in such conditions, Olsen wasn't sure what he'd found. He picked up the scores of half-centimeter-sized fossil bits and brought them back to Horner and Baird. He watched in amazement as "they pieced the thing together—it was a complete neck vertebra, and they used up all the pieces."

Olsen was far more interested in the geology of Wassons Bluff than in the mammal-like reptile he and then Harvard graduate student Neil Shubin had found along the beaches there in 1984. Olsen had recognized the potential of the site even before the finds were made, particularly the opportunity to fix the time and habitat of fossils with exactness unparalleled in dinosaur studies.

Like Olsen, his crew was expert, although dinosaurs were but a small portion of their research interests. In their minds, and those of most other paleontologists today, studying only dinosaurs would be an unreasonably limited specialization. To them, dinosaurs matter, but equally with all animal and plant inhabitants of their time. Cumulatively, these creatures tell the most important stories in science, sagas of the shape and pace of evolution and environmental change.

Olsen's chief collaborator, Neil Shubin, initiated their cooperative proposal to the National Geographic Society for funding in 1985. Barely thirty, bearded and cherubic, Shubin is not a dinosaur paleontologist but a structural morphologist, a scientist who studies the evolution of anatomical features in a variety of creatures. Shubin is now an assistant professor of biology at the University of Pennsylvania. He's looking closely at the Nova Scotian dinosaurs he's unearthed with an eye toward their structural precursors among earlier animals in the fossil record. "I want to know what changed and how," says Shubin.

Another regular coworker in Nova Scotia is Hans-Dieter Sues. Like Shubin he is a Harvard paleontology Ph.D. and an outstanding scholar. But Sues subsisted by fellowship (at the Smithsonian) for several years before landing a position in 1992 as a curator of paleontology at the Royal Ontario Museum in Toronto. Sues became a dinosaur fancier early in life, at the age of four, in Düsseldorf. "My mother bought a book on dinosaurs for me. I decided that was for me. She thought I'd outgrow it." Sues hasn't, though now he's interested in "faunas as a whole, patterns of mass extinction and other global events, the struggle between adaptation and constraint, revealed by systematics, taxonomy, sedimentology, stratigraphy. The great paleontologists are trying to solve problems, not just add another data point."

The other crew member is William Amaral, a veteran Harvard Univer-

Patrolling the beaches beneath Wassons Bluff, Nova Scotia, for fragments of dinosaur bone from a pivotal time in dinosaur evolution are, left to right, paleontologists Neil Shubin, University of Pennsylvania, and Hans-Dieter Sues, Royal Ontario Museum, and preparator William Amaral, Harvard University. [Photo by author]

sity preparator, renowned as the most skilled and patient in his field. He once spent two years cleaning three walnut-sized mammal skulls.

Olsen is the chief scientist in this group, for "this man knows everything about the Triassic-Jurassic boundary. And if there was something he didn't know, he could guess it," says Johns Hopkins paleontologist David Weishampel.

The Newark Supergroup underlies the eastern United States and occasionally breaks the surface, as it does at Wassons Bluff. It is exposed also at Olsen and Sues's latest site, a low hillside just outside Richmond, Virginia. In those Triassic sediments Olsen and Sues found the jaw of perhaps the most immediate ancestor of modern mammals yet known, some 225 million years old.

Olsen began studying the Newark Supergroup in the backyards and stone quarries of central New Jersey, where he first dug fossils at the age of eleven. By junior high school he was traveling (by bus) on his own around the heavily developed Garden State in search of fossils. Says Padian, "Paul went everywhere looking for rock outcrops. He's been thrown off people's property, walked around behind shopping malls." (Padian and Olsen spent one summer looking for fossil fish in an outcrop by a ventilator shaft for the Lincoln Tunnel.)

In his youth, Olsen's favorite site was a quarry near his house in what is now Walter Kidde Dinosaur Park in Roseland, New Jersey. There, at fourteen, he found dinosaur footprints and spent three years excavating. As a Yale undergraduate, Olsen seemed intent on not getting a degree. Says Padian, "Paul had pretty much dropped out." Olsen was no disillusioned or unmotivated student. Rather, as Padian recalls, "he stayed in the basement of the Peabody Museum and worked."

What Olsen worked on in college was the much-neglected science of dinosaur footprints, even running crocodiles and rheas to observe the pattern of their prints. He continues to dabble in footprint research while concentrating on unlocking the rock and fossil mysteries of the Newark Supergroup. Before Olsen, the Newark Supergroup was a little-known and lightly regarded sequence of rocks. "It was terribly studied," says Olsen, shaking his head in disbelief. "One geologist called it a monotonous series of barren red beds."

The Supergroup

Thanks to Olsen, the Newark Supergroup is now known to have dozens of black and gray sediments, abundant in fossils. Countless outcrops appear from Nova Scotia to South Carolina in a formation that continues, beneath today's coastal plain, north to Newfoundland and south to Florida. Matching basins to the Newark Supergroup's are found in Morocco and the Iberian Peninsula on the other side of the spreading rift that eventually became the Atlantic Ocean. What's more, the Newark Supergroup's geological, paleontological, and environmental history is now known in sometimes exquisite detail, thanks largely to Olsen. With that understanding have come new clues to early dinosaurs, and the catastrophic changes in their environment.

The Newark Supergroup preserves sediments from parts of a rift along a fault line. The sediments were laid down in large lakes, most of them shallow, but some more than five hundred feet deep, that formed when large blocks of crust along the rift separated, leaving holes that filled with rain. Some of the ancient lakes were large indeed—some thousands or tens of thousands of square miles in area.

All the lakes began collecting sediments some 235 million years ago and continued to do so until the end of the Newark episode 185 million years ago, when the lakes disappeared into a new and widening Atlantic Ocean. There were no known dinosaurs when the Newark Supergroup began forming, but by the time it stopped building, dinosaurs ruled the earth.

Since 1970, Olsen has been traveling the North American east coast, examining all the rift basins. In their ancient lake deposits he has detected patterns of environmental change that occurred with clockwork regularity, regulated by the earth's own changing movement. Olsen discovered that "the lakes are incredibly sensitive to amounts of precipitation." Repeating colored bands from black shale to red mudstone mark outcrops of the Newark Supergroup, reflecting cycles in precipitation. When rainfall increased, the lakes deepened, leaving fish fossils within its black shale deposits from fish that died in oxygen-starved lake bottoms. Massive mudstones, full of fossil plant roots, record a verdant environment at times when water levels were rising. From high to low to high again, the lakes drained and filled in 21,000-year cycles. The timing of those cycles was affected by no fewer than three astronomical cycles (known as Milankovitch cycles). The most influential was the shifting earth axis. The Greeks and Babylonians had observed the changing position of the North Star over hundreds of years, and how the dates when day length equaled night changed from year to year—the procession of the equinoxes. Those changes were quantified by Kepler and Newton's description of the earth's orbit. The day-length changes reflected the earth's changing aspect toward the sun.

Now the Northern Hemisphere is closer to the sun in its winter than is the Southern Hemisphere in its winter. But in ten thousand years the southern half of earth will be closer in winter. The climate of the northern lands and oceans will be colder in winter. This shift in earth orbit, according to Olsen, has always had "a profound effect on climate."

A second climatic cause for the fluctuating levels of Newark Supergroup lakes was the obliquity cycle. Now the planet is cocked at a 23.5-degree angle. But that angle shifts by several degrees on a 41,000-year cycle. And, as Olsen notes, "the greater the tilt, the greater the contrast in climates."

Yet another cycle to be reckoned with is the changing shape of the earth's orbit. On a 100,000-year cycle, the path of the planet shifts from a 6 percent ellipse to a near-perfect circle and back to ellipse again. In elliptical orbit climate is, as one would expect, more varied throughout the year than in times of circular orbit.

Olsen has factored these and other astronomical changes—the effect of lunar pull on the equatorial bulge to make the earth wobble, eccentricities in the orbits of Jupiter and Saturn that vary the elliptical orbit of the earth in, respectively, 2,000,000- and 400,000-year harmonics—into his analysis of the climates of eastern North America at the time of the great extinction of 200 million years ago. "All these changes affect the amount of sunlight and change its distribution," says Olsen. "Until the last twenty years their influence was thought to be small. But the great ice ages, monsoons, and major droughts in East Africa have been linked to these cycles."

Monsoons, sweeping across the supercontinent of Pangaea, are what filled the lakes of the eastern American rift valley. Knowing the cyclic pattern of the lakes and how they left their mark on the Newark Supergroup, Olsen is able to date specific fossils and environmental events with an accuracy new to such ancient sediments. Anywhere else in the world, it is impossible to measure the age of a fossil, whether by chemical dating or its association with well-dated marine fossils or mammal teeth or pollen grains, to any more exactitude than the nearest one hundred thousand years. In the Newark Supergroup, Olsen can pinpoint the age of fossils to well within a specific 21,000-year cycle. This precision makes possible far more detailed judgments about the manner and speed of evolution as well as the role of the environment in those changes.

Thanks in large measure to Olsen's years of work, the Newark Supergroup has become the best place on Earth to figure the age of Triassic-Jurassic rocks, plants, and animals. The long, precisely repeating patterns of sediment accumulation, the abundant record of pollen, and of vertebrate fossils and footprints, can all be correlated with geochemically derived dates from its basalts.

Even specific dinosaur footprints can be used in dating. Footprints were left only at the midpoints of the lake-making cycle when mud flats were exposed—ideal conditions for preserving dinosaur footprints. Therefore, those footprints can be used to identify the midpoints within the 21,000-year cycle of Triassic-Jurassic lake formation.

Another highly accurate marker within the Supergroup came from volcanic activity along the length of the formation that caused a slow but substantial eruption. The extruded molten rock made thick lava flows that settled into a lava lake more than three hundred feet thick. The basalt cones at Wassons Bluff formed from those eruptions. Unlike most dinosaur-bearing rock, which is formed from the slow buildup of sediment, volcanic rock can be dated geochemically with more than 99 percent accuracy to the moment when it was created.

Geochemical dating of volcanic rock involves measurements of radioactive trace elements. The ratio of one unstable element to its product of radioactive decay indicates the time taken by that transformation, as we know the half-life of these elements. (Such tests are highly sensitive. In looking for a stable trace element, iridium, researchers at Walter Alvarez's University of California/Berkeley laboratory obtained strangely high amounts of the element in a rock sample, throwing off his dates by millions of years. Iridium is found in platinum. A researcher's wedding band had brushed against the rock.)

Colleagues of Olsen's at the United States Geological Survey and the Royal Ontario Museum devised a variety of chemical dating methods that suggest in sum that the basalts at Wassons Bluff are 201 million years old (to an accuracy level of plus or minus 2 million years). These dates place the rock at Wassons Bluff right at the boundary of the Triassic and Jurassic periods, according to Olsen's measurements. (Other scientists have placed the boundary at 208 million years.)

Olsen's dating has shortened the reign of the dinosaurs. With revised dating, dinosaurs now seem uncommon and far from dominant before 200 million years ago. The fauna of Wassons Bluff represent the first land animals from the beginning of the Jurassic period. Dinosaurs are a significant element of a community that includes predominantly crocodilelike land animals, as well as lizardlike creatures and some mammal-like reptiles. All of these holdovers from the Late Triassic were small, dinosaurs included, typically less than three feet long.

But if little dinosaurs and other small creatures were surviving in Nova Scotia and around the globe, other larger Triassic animals were disappearing. Amphibious, mammal-like reptiles and many archosaurs died out. The calamities that transpired at the end of the Triassic, 200 million years ago, remain mysterious. Why small land animals survived when many larger ones died out is uncertain. Clearly many organisms became extinct between the Late Triassic and the Early Jurassic. But some paleontologists have argued that the marine and land extinctions were not synchronous. And Mike Benton, as mentioned earlier, has reasoned that many of the dominant Triassic land animals went extinct well before the boundary.

Olsen doesn't buy Benton's reasoning. "There's no evidence the extinctions he's talking about were sudden. If you plot family diversity through the Triassic, it goes up in the Late Triassic to the boundary. Then about half of the kinds of creatures don't make it across." But Olsen admits, "It's not clear just where it drops off. It could be six hundred thousand years; it could be just at the boundary."

Either way, by geological standards, the extinctions at the boundary happened fast, fast enough to fit with another of Olsen's astronomical explanations for earthly change: "I don't love slow extinction. I like asteroids and sudden extinction better.'

In with a Bang

If there was a sudden extinction 200 million years ago, it doesn't appear to have catastrophically affected the small dinosaurs. Although the extinction at the end of the Cretaceous, 65 million years ago, saw the end of all dinosaurs, smaller versions of large plant-eating dinosaurs from the Triassic of Europe have been found from early in the Jurassic of Africa, Asia, and South America, prosauropod dinosaurs in particular. Their survival suggests that their food sources, plants up to thirteen feet high, also endured. Small animals, little dinosaurs included, would be more likely to survive an extinction catastrophe because they required less food, reproduced more quickly, and could have hidden in burrows, according to Olsen.

At Wassons Bluff, Olsen had the opportunity to test whether his sudden extinction theory was true. Pollen findings appear to support the idea of a disruption of the terrestrial ecosystem. A substantial increase in the amount of pollen of one plant group occurs after the boundary. These pollen grains belong to conifers, from small bushes to trees of mature-forest height. Their small pointed leaves suggest these plants were adapted for dry climates. Whether gradually or by some swift extraterrestrial trigger, some climatic change killed off the prevailing flora of Nova Scotia, and conifers moved in as plant colonists. The plant-eating reptiles that became extinct at the time could have been those that did not adapt to the changeover to a diet of conifers.

If the environmental shift was sudden, the event that caused this shake-up of plant and animal communities might, Olsen reasoned, be recorded. He thinks that record exists at Manicouagan, Quebec.

Manicouagan is a ring-shaped lake, some fifty miles in diameter, five hundred miles from Wassons Bluff. For decades geologists have speculated the lake lay in a crater made by the collision of an extraterrestrial object with the earth some 200 million years ago. To Olsen the supporting evidence is now incontrovertible: "You've got a large circular structure whose interior is completely melted and rehardened rock. The rock underneath and surrounding it is 1.6 billion years old, but inside the crater the radiometric date is 200 million years old. Around the edges of the melted and remixed rock is terribly shattered rock. There are shatter cones [patterns of fracture in rock] radiating away from the circular structure."

The science of meteorite detection is scarcely older than that of dinosaur identification. It was not until 1803, when two thousand meteorites fell in a hot shower on the village of L'Aigle, France, that scientists acknowledged the reality of rocks falling from space. The search for impact craters has quickened considerably in the past decade as physicists seek to test notions of meteorite-caused dinosaur extinction. Yet only 120 craters on earth are well documented, none larger than ninety miles across. Solid evidence linking an extraterrestrial impact with Triassic-Jurassic boundary extinctions has been hard to come by. At the K-T (Cretaceous-Tertiary) boundary, 65 million years ago, an increased concentration of iridium can be found at many localities worldwide. This is not the case at Manicouagan. Iridium does not naturally occur in the earth's crust in appreciable quantities, so an increased level in a rock sample is believed by most physicists to be of extraterrestrial origin. It might perhaps even be related, as the late Nobel-laureate physicist Luis Alvarez postulated, to a fiery finale for the dinosaurs.

But the absence of unusual concentrations of iridium at Manicouagan does not preclude an asteroidal impact, even one of earth-changing proportions. There are two types of meteorite, as Olsen explains. One contains iridium, but the other, composed of meteoritic basalts, resembles more closely the earth's crust, and like it, this meteor does not naturally contain much iridium.

Manicouagan has something else only asteroids can bring in significant amounts: shocked quartz. Under extreme pressure, greater even than that generated by volcanoes (in most geologists' view), quartz grains take on a peculiarly striated appearance. "Normally they look like pages in a book," says Neil Shubin, "but an asteroid or a nuclear explosion blows open the crystal lattice, making a herringbone or checkerboard pattern."

Shubin, Olsen, and three others collected shocked quartz among the samples they took to redate the Manicouagan crater, in a foray that nearly became what Shubin calls "a minor extinction event." They piled into a twenty-foot cargo canoe with a thousand pounds of equipment and set off across the cold lake. Midway, the engine died, water poured in over the bow, and only frantic rowing and bailing got them to shore.

At Wassons Bluff, the group began searching through sediment of the appropriate age for shocked quartz. To Olsen, it is plausible that a crash hundreds of miles away altered rock crystals on the shore of Nova Scotia. "To make a crater the size of Manicouagan we're talking of a fireball fifteen hundred miles in diameter. This was something that stretched halfway across North America," and, as he said in *National Geographic* magazine, "scorched everything down to New Jersey."

Determining whether the impact of an asteroid at Manicouagan created shocked quartz at Wassons Bluff is but one step up the stairway to linking the event to a catastrophic end to the Triassic. Additionally, the "absolute" chemical dates for the rocks at Manicouagan, which are 213 million years old (plus or minus 500,000 years), must correlate with the dates obtained for Wassons Bluff—201 million years ago (plus or minus 1 million years).

These dates don't come close. But, says Olsen, the dates for Manicouagan "are vague and equivocal. Essentially, I don't believe them." And the methods used were different from those at Wassons Bluff. "It's like comparing apples and oranges," says Olsen, who's more confident of the Wassons Bluff age. "It's nailed at 201 million years. It matches many, many dates taken from fissures from South Carolina to Nova Scotia."

Olsen plans to further test Manicouagan and Wassons Bluff samples. Even if a connection can be found that makes the dinosaurs of Wassons Bluff into near-term survivors of a Manicouagan-centered impact, it would be difficult to establish if Manicouagan was the shock felt around the world. Nor could it be demonstrated the blast was sufficient to trigger a global Nuclear Winter or other drastic environmental change fatal to many species. As Padian has said, "It's quite possible that the extinction along the east coast of North America occurred fairly quickly. But whether that has general applications around the world is still an open question." He suggests the intense volcanic activity and extensive faulting evident in the basalts of Wassons Bluff might themselves have caused extinctions of the animal and plant populations of the Newark Supergroup.

It is difficult, too, to understand why dinosaurs survived such a disaster, only to perish from another 135 million years later. Perhaps, following Olsen's reasoning, the dinosaurs had grown too large to survive the asteroid impact of 65 million years ago.

As with so many of the often-asked questions of dinosaur life—what did they look like, how were they created, how did they die—the how and why of the dinosaurs' domination of the world are nearly unanswerable. In the contingent view of evolution promulgated by Stephen Jay Gould, there is no explanation but chance: "Wind back life's tape to the dawn of time and let it play again—and you will never get humans a second time." The same holds for dinosaurs. However successful they were (and our own dominion comes nowhere close to the 130-million-year dinosaur reign), the dinosaurs were accidents.

Even if we can never explain the dinosaurs' triumph, nor their fall, we can come nearer to understanding their lives and times. To do so requires more tools than a pick and shovel, more talents than an eye for bones. One place at a significant moment in dinosaur time is now uncommonly well-understood, thanks to the talents of Paul Olsen. But that evidence—of several relatively small dinosaurs thriving at the precise time of marked change in the land, the climate, and the vegetation, and possibly coincident with a major asteroidal impact—is still open to interpretation.

Olsen acknowledges he's far from substantiating a temporal, regional, or global link between the Manicouagan impact and sudden extinctions worldwide. "So far," he says, "things don't line up. We've got a long way to go before saying whether this extinction was abrupt or gradual."

But Olsen holds to his asteroid hypothesis. And if Manicouagan doesn't pan out, "there is a very large crater in the Soviet Union called Puchezh-Katunki," he says with enthusiasm. "It's even bigger than Manicouagan. It could be Early Triassic to Late Jurassic. There's only one date for it, 183 million years ago, and it's no good. But I haven't had the time to look into it. And something else strange—there was a giant impact on Mars from which ejecta is still coming down to earth. Turns out that also happened 200 million years ago."

Olsen is excited but hardly swayed by the possibilities these new leads offer. "I'm not one of those crater-crazy people. But I don't know what to make of it. Clearly, this was a busy time."

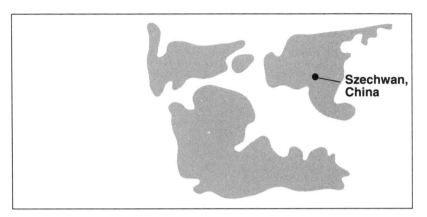

Pangaea starts to split, 188–163m. yrs. ago

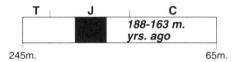

T	J	C
		188-163 m. yrs. ago

245m.   65m.

AGE OF THE DINOSAURS

Triassic Jurassic Cretaceous

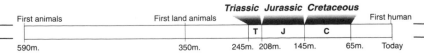

First animals   First land animals   First human

			T	J	C	

590m.   350m.   245m. 208m.   145m.   65m.   Today

The Dragons of Dashanpu

Huge ferns crackle and crash to the fertile valley floor, crushed beneath the massive footfalls of advancing giants. Above, monstrously long-necked titans stretch to the upper edges of palm trees. At the top of the canopy, visible only to a circling pterosaur, the oddly small, blocky heads of the beasts thrust upward. Their gaping mouths reveal rows of spoon-shaped teeth, gnashing clusters of leaves.

Nearby, in this lush, moist land, other four-legged behemoths, short-necked and thick-toothed, tear away at the verdant understory. Tiny foragers sprint on spindly hind legs, skirting trees and the legs of the placid diners. Noisily crunching on ground plants are a pair of bulky beasts, adorned with odd and menacing plates and spikes down their backs. Short horns over the eyes distinguish the male from its mate.

The pastoral calm is broken by a menacing roar—a huge and nimble predator, twenty feet high, rushes in. Cries of alarm fill the valley as small animals flee in panic. The earth shakes with each footfall of the still-hidden menace. Among the browsing giants, adults shield their young in the center of a clearing. With sudden savagery, the hunter charges, flashing huge dagger teeth . . .

A scene from life in Szechwan, China, 160 million years ago. It is a moment in time from the heyday of the dinosaurs. And until just a few years ago, this was a time, and a menagerie of bizarre beasts, entirely unknown to us.

In a Greek delicatessen in a one-horse Utah desert town, a ponytailed New Jersey–born dinosaur paleontologist in a cowboy hat is looking at photographs. He is also sharing a pastrami sandwich with his Chinese counterpart, a diminutive man whose face is obscured by thick glasses and a blue Seattle Mariners baseball cap. The one, perhaps only, common interest of the diners, Bob Bakker and Xiao Xi Wu, is what they are examining in the blurry color images—dinosaurs.

The pictures are Chinese, taken of the world's finest dinosaur fossil locality—the Dashanpu quarry in Zigong, Szechwan, where Xiao worked for three years. Thanks to recent well-attended exhibitions of Chinese dinosaurs in Tokyo (which drew 1 million visitors), Seattle, Cardiff, New York, Boston, Los Angeles, Sydney, Stockholm, and London, it is now no secret to the general public that China is wonderfully rich in dinosaurs.

But even to professional paleontologists, the extent of the best and most distinctive Chinese dig was little understood when Xiao came to America in June of 1988 to open the exhibit of Szechwan dinosaurs in Seattle.

Excavations in the purple mud of Dashanpu began in 1979, and reports began appearing in Chinese journals in 1984. But no specimens, and few photographs of the quarry's spectacular contents, had been viewed by Western paleontologists before Xiao's visit, save by those few who had journeyed to southern China.

Dinosaurs little-known or entirely unknown elsewhere were found dazzlingly complete at Dashanpu, including skeletons of a thirty-five-foot-long giant, an animal dubbed *Shunosaurus,* the name deriving in part from "Shuo," the old Chinese name for Szechwan. There are less than a score of skulls of any of the many kinds of huge sauropods, and Dashanpu has three in outstanding condition. When found, the Dashanpu dinosaurs looked like museum-mounted specimens that somehow toppled over, intact, into the purple mud.

The significance of these skeletons goes far beyond their remarkable state of preservation. In sum, they constitute one of the most complete records of dinosaurs from any era, in this case from an eventful era that was all but unknown until the Szechwan exploration: the heart of dinosaur time, the middle of the Jurassic period. To Bakker, "the Early Jurassic looks like the Late Triassic, not a lot happening. But in the middle, all hell breaks loose. But you have only fragments from Madagascar, Australia, until Zigong. Here's a whole fauna twenty-five million years earlier than the next."

A fertile, sometimes sweltering valley now, Szechwan was a rich tropical environment in the middle of dinosaur times. Vegetation was luxuriant. A leafy understory was formed by palmlike cycads that had long stiff leaves and pineapple-ish diamond patterns on their bark. The understory was filled out by ginkgo trees (still a common ornamental tree) and ferns, horsetails, bushes, trees, and club mosses with thin stems and narrow leaves. The dominant trees forming the canopy layer were needle-leafed conifers. This environment predominated around the world at the time, as pollen and fossils from England and France indicate. A huge petrified tree stump in the Zigong Museum testifies to the tropical forest that was once Szechwan.

Through this forest roamed unfamiliar dinosaurs great and small. Grandest of all were long-necked sauropods nearly seventy feet long. Their vast columnar legs shook the ground as they plodded through the forest. They paused to tear at palm treetops, browsing with spoon-shaped teeth set in relatively tiny, deep skulls. Each huge herbivore's long tail sported a club at the tip. This would have been a useful weapon when wielded to fend off predators such as *Gasosaurus,* a bulky, two-legged, sharp-toothed hunter nearly fifteen feet long.

Shunosaurus was half as large as the giant plant-eaters, a mere thirty to thirty-five feet long. Like the behemoths, it also ambled through the Szechwan forest, likely stretching its relatively short neck to feed on lower-level foliage. Yet another four-legged plant-eater, half again as

Gasosaurus *[John Sibbick]*

small as *Shunosaurus,* named *Datousaurus,* foraged here. Its long tail ended in an even more frightening weapon, four short spikes as well as a bony club.

A pair of stout four-legged dinosaurs nibbled on low-growing plants. Adorned from head to tail with a double row of spikes, these creatures grew to thirteen feet in length. One, perhaps the male, had short horns over his eyes as well. Other plant-eaters, slight, quick, and no bigger than a child, darted through the underbrush on their hind legs.

In their staggering completeness, this community of newfound dinosaurs filled a gaping hole in our understanding of dinosaur evolution. For two groups of giants familiar from later eras, the giant sauropods and the plate-backed stegosaurs, Dashanpu provided especially surprising answers about their early appearance.

The Dashanpu quarry sits on a low hillside eight hours by oxcart-filled roads from the nearest metropolis, the city of Chengdu. Zigong, the town adjoining the Dashanpu quarry, is a hamlet of a few thousand, seven miles southwest of the quarry, and a Chinese tourist attraction. More than 1 million people, largely from within the highly populous Szechwan province, come to Zigong's spring festival to buy and sell dragon lamps—lanterns of colored silk or cloth, decorated with or shaped into dragon images.

Since 1987, the Dashanpu dinosaur excavation has formed the ground floor of the Zigong Dinosaur Museum, a large concrete structure, sparsely decorated but elegant nonetheless, thanks to the graceful, peculiar forms of its giant mounted dinosaurs. Huge skeletons are arrayed in the cavernous upper gallery, while the few visitors, mostly schoolchildren, parade past and look down upon the quarry itself. The site of this massive bone bed, still chock-full of many recognizable fossils, hard by the huge animals that have emerged from it, makes for one of the most exciting dinosaur displays in the world.

Here as elsewhere in China, dinosaurs have long been treasured. The Chinese term for dinosaurs, *kong long,* means "dragon bones." Dinosaur fossils are still widely taken for physical remains of the mythical beast central to Chinese philosophy (1988 was the Year of the Dragon in China). The *lei wen,* the dragon and cloud motif so prevalent in Chinese art, can be seen on the earliest ceremonial bronzes known, elegant vessels from the Shang dynasty of over three thousand years ago.

Dinosaur bones—or dragon bones, as they were believed to be—have been dug for centuries in China, to be ground up for traditional medicines and poultices to treat burns. A Chinese sage of the twelfth century, Chu Hsi, may have been the first man anywhere to note that other such fossils were the remains of ancient life: "I have seen on high mountains conchs and oyster shells, often embedded in the rocks. These rocks in ancient times were earth or mud, and the conch and oysters lived in water. Subsequently, everything that was at the bottom came to be at the top, and what was originally soft became solid and hard. One should meditate deeply on such matters, for these facts can be verified."

The first Chinese dinosaur bones known, though not immediately

The stunning dinosaur finds of Dashanpu are remounted in Asia's largest fossil museum in Zigong, China. [Photo courtesy New China Pictures Agency]

Omeisaurus *(top)* *and* Huayangosaurus *(bottom)* *[John Sibbick]*

verified as such, were acquired by a Russian colonel from fishermen along the Black Dragon River in northeastern China in 1907. They were mistaken for nearly a decade for Siberian mammoth bones, until Russian scientists returned to the site and found the bones of a large duckbilled dinosaur. This they named *Mandschurosaurus,* the first in a long litany of Chinese dinosaurs named for the tongue-tying localities in which they were first uncovered. An American geologist reported dinosaur bones from central China on his 1913–1915 geological survey of Szechwan. And a German researcher, Dr. Otto Zdansky, led the first dinosaur dig in China, in Shantung Province in 1923, and found a robust four-legged browser, a sauropod dinosaur thirty to thirty-five feet long. Americans and Germans, Swedes and French, followed to dig dinosaurs across China in the 1920s and 1930s.

Dinosaur paleontology in China didn't become a Chinese science until a scholarly young paleontologist, Yang Zhong-jian (better known as C. C. Young to Westerners), returned from Munich with his doctorate in the late 1920s. For three decades thereafter, Yang led excavations throughout his country and turned up many new genera of dinosaurs.

Among them were many peculiar forms of the huge sauropods. In 1937, Yang found a thirty-five-foot-long sauropod in China's northwest. Soon after, in Szechwan in 1939, Yang, in the company of University of California/Berkeley paleontologist Charles Camp, came upon a sauropod estimated to be some forty feet long. This animal, *Omeisaurus,* was the first of the distinctively long-necked Chinese sauropods to be unearthed. These giraffe-ish dinosaurs evolved light and mobile necks, held upright (according to the upward bend in their backbones). These necks were as long, or longer, than the rest of their bodies. But since Yang located only four neck vertebrae, he had no inkling of this anatomical peculiarity.

Yang did have the good fortune to have been a school chum of Mao Tse-tung's, and when the People's Republic was created in 1949, he became director of the new national agency for the study of ancient animal life, the Institute of Vertebrate Paleontology and Paleoanthropology (then called the Laboratory of Vertebrate Paleontology) in Beijing.

But Yang continued to miss out on the peculiar anatomy of some

Chinese dinosaurs after the Maoist revolution. In 1952, tipped off by construction workers who'd come across the fossils while working near the Mamenchi Ferry on the Yangtze in Szechwan, Yang excavated a ludicrously long-necked sauropod, an animal named *Mamenchisaurus* (for the ferry stop, with the species name *constructus* to honor the workers who found it). This animal had the longest neck of any creature ever known on earth, a preposterously long forty-foot tube of flesh, muscle, and bone. But on this first specimen the neck bones had been crushed, leading Yang to reconstruct it with a wrongly foreshortened neck.

A petroleum expedition helped set Yang straight. Oil workers in Szechwan located a mamenchisaur in 1957, and the Qongqing Museum unearthed it all in three months. At more than seventy feet long it was the longest dinosaur ever discovered in Asia.

Yang's most celebrated find, of another sort of dinosaur entirely, was made in 1950 on the coastline of Shantung Province. There Yang came upon the bones of a huge twenty-five-foot-long duckbill, *Tsingtaosaurus,* which he reconstructed with an odd, unicorn spike protruding from its skull. It looked somewhat like a duckbill common to North America and Asia, but unlike other duckbills, *Tsingtaosaurus*'s narrow crest pointed forward, not backward, at least in Yang's reconstruction. In its present state, the duckbill's skull is encrusted with so much plaster that only an X ray could determine if Yang has reattached the spike properly.

Yang's successor as the emperor of Chinese dinosaur paleontology, Dong Zhiming, was a nature-loving schoolboy in Tsingtao when Yang found his unicorn dinosaur. *"Tsingtaosaurus* was found close by when I was in middle school, and my school visited [the site]," Dong recalls. His appetite for paleontology whetted, Dong went to Fudan University to study vertebrate biology before coming to the IVPP in 1962 to work for Yang.

Dong was soon digging for dinosaurs all over China. In nearly three decades since as the IVPP's chief dinosaur scientist, Dong says he has excavated more dinosaur bones than anyone else in paleontological history. He excavates six months a year, in every corner of China, leading his wife to complain, "All he does is dream about dinosaurs."

The world's most prolific dinosaur hunter, Chinese paleontologist Dong Zhiming, seen here on expedition in the Gobi Desert. [Photo by author]

Dong is no dreamer. He is a bulky, voluble, headstrong man with a ready smile and high-pitched laugh. His openness sets him apart from his Chinese colleagues and puts his friend Xiao Xi Wu in mind of Bakker. "They are both bright, friendly, unkempt, dirty. Dong is your friend right away." But to New Mexico paleontologist Spencer Lucas, a coworker of Dong's in 1984, he is "an egomaniac. He sees himself as China's gift to dinosaurs." The opinions of Westerners who've worked at length with Dong fall somewhere in between these perspectives. Some Western scientists criticize his stubbornness and say his research lacks sophisticated analysis; others praise his enterprise and intellectual curiosity. Still, even his harshest critic, Spencer Lucas, admits, "Dong is very smart for a guy with no advanced degree. He's a good organizer and he gets the bones out of the ground. Evolutionary mode and tempo, paleoecology, are beyond him, but he doesn't have the literature, the tradition." (I found him charming, accessible, and savvy. And I am personally indebted to him and his red-tape-cutting skills for giving me the unprecedented journalistic opportunity of observing Chinese digs.) However he is viewed, Dong is unquestionably the central link between China's rich history of dinosaur exploration and its even more promising future.

In addition to supervising the Zigong digs and the Chinese-Canadian Dinosaur Project, the largest dinosaur expedition of recent times, Dong has excavated dinosaurs in nearly every province of China. These dinosaurs encompass nearly every era of dinosaur history except the earliest, in the Triassic of 225 million years ago. As Dong has written, with only slight chauvinistic excess, "China is the only country in the world where enough of the pages [of dinosaur life] have been preserved intact to show what was happening throughout their history."

Dong has led four expeditions to the Junggar and Turpan basins in the northwest of China and in far northeastern Heilongjiang. In Shantung Province, Dong excavated the largest duckbilled dinosaur ever found (*Shantungosaurus giganteus*)—fifty feet long and nine tons in weight. In the Nanshiung Basin of far southern China, he found one of the smallest duckbills (*Microhadrosaurus*, only eight feet long). And in Szechwan, Dong found the nearly complete skeleton of a twenty-five-foot-long predator, *Yangchuanosaurus*, the best-preserved carnivorous dinosaur ever found in China.

These are but a few of the tens of bone, egg, and footprint localities identified throughout China, nearly all of them investigated by Dong.

While China is rich in dinosaurs and has climates that make many of the dinosaur sites accessible throughout much of the year, Dong and his fellow Chinese dinosaur paleontologists labor under enormous hardships. Funds and equipment are hard to come by for any Chinese science, and the IVPP, according to its paleontologists, does not enjoy high priority among the various disciplines within the national Academia Sinica. Yang's friendship with Mao did help shield the IVPP and its staff members from the harshest deprivations of the Cultural Revolution (at least one IVPP geologist died in the turmoil, according to Lucas). "All scientists were sent into the country for reeducation," recalls Dong. Smiling, he adds, "I was already in the country, so I was okay." His friend, Sankar Chatterjee (who, like Lucas, worked with Dong in 1984), says Dong is downplaying his difficulties. "He was sent to the farms for many years."

Xiao Xi Wu and colleagues from the Qongqing Museum fared even less well during the Cultural Revolution, but were still fortunate to be spared the internment or execution faced by many academics. Rather, Xiao was confined to virtual house arrest for more than two years when his museum was closed.

The IVPP may have fared better than other institutions in the Cultural Revolution, but it suffered greatly nonetheless. As Lucas noted on his 1984 visit to Beijing, "A lot disappeared from the IVPP during the Cultural Revolution. Half the library was destroyed. The type specimen for Yang's original description of *Omeisaurus* disappeared."

In the field, the laboratory, and in publications, Chinese paleontology suffers to this day for the nation's past political orthodoxy and isolation and its continuing economic difficulties. Westerners who come to China aware of all this are nonetheless unhappily surprised by the primitive scientific facilities in China.

When I toured the spacious Museum of Natural History in Beijing in 1988, so many of its abundant dinosaur skeletons had been removed for revenue-generating foreign display that the ground floor was half-vacant, and the entire second story converted to a furniture showroom.

On the outskirts of Beijing, at IVPP headquarters, paint peels off the walls of the offices. Fossil workers work under bare light bulbs in open-sided sheds and live, as Dong does, in modest communal housing on-site. Visiting paleontologist Paul Sereno found that "it was hard to get much done at the IVPP, hard to get anything, even borrow a lamp." The cataloguing of fossils is in shambles, and many researchers simply keep their study specimens beneath their desks. Techniques are primitive, and skeletons are many times a sorry jumble of jerry-rigged parts and plaster inventions.

Were all the Chinese dinosaur skeletons to return from their tours, no public facility is sufficiently large to display them. A new IVPP headquarters, similar in size to the spacious Royal Tyrrell Museum, is now planned for downtown Beijing, near the zoo and convention center, at a cost of $4 million. Ground-breaking was scheduled for June 1989, but by spring only half the projected budget had been pledged by the national science academy. The building was still unfinished as of early 1992.

But, Sereno notes, given its meager resources, the achievements of the IVPP are stupendous. The institution houses more than 120,000 fossils. It supports 160 scientists and 260 employees, a staff maintains its facilities, however humble, and it sponsors field research throughout the country—all on a budget of less than $250,000 a year.

In the field, the Chinese dig with a level of sophistication Western scientists compare to the Cope and Marsh Dinosaur Gold Rush of the late nineteenth century. Quarries are excavated with dynamite, bones pried from the ground without recording their locations or the flora and fauna alongside. Those in charge, including Dong, have little concept of evolutionary or ecological theory, few techniques for careful excavation, and little equipment to carry out the work.

For Chinese paleontology the hope for progress lies not with Dong or his contemporary Xiao, schooled at Xian University, but with the next generation of paleontologists, among them Western-trained IVPP students, such as Li Ming, who has studied at Brigham Young University, and Yu Chao, who works with Sereno at the University of Chicago and who accompanied him on his dig in Ischigualasto, Argentina (see Chapter 2).

Digging Dashanpu

That dinosaurs would be turned up at the Dashanpu quarry was hardly surprising to anyone even vaguely familiar with Chinese dinosaur paleontology. Szechwan has yielded, over the last half century, the longest continuous record of dinosaurs from their central era of any region in the world, a happy accident of its geology. The name Szechwan itself means four rivers converging into a plain. Szechwan stood from 195 to 140 million years ago at the headwaters of a vast inland lake that is now the source of the Yangtze. A sedimentary basin the size of Rhode Island, the Szechwan plain filled steadily to produce a continuous sequence of soft mudstone—hugging the fossilized bones of many dinosaurs—for some 70 million years, from the Late Triassic (215 million years ago) to the Jurassic-Cretaceous boundary (145 million years ago).

Chinese schoolchildren were paraded past the world's most spectacular dinosaur quarry in Dashanpu, Szechwan, as workers excavated the nearly intact skeletons. [Photo courtesy New China Pictures]

According to *Dinosaurs From China,* a handsome, largely pictorial book coauthored by Dong and British Museum paleontologist Angela Milner, "the deep red mud and silt of the lakeshore was the perfect medium for preserving bone." This Szechwan Basin mudstone is two miles thick in some places within the province. It has proven rich, in spots, in both oil and gas.

In 1979, a Chinese energy-exploration unit came to Dashanpu to dig for natural gas, and to build a factory for the construction of trucks and buses. Halfway down a steep hill, they struck dinosaurs. By law, the construction workers had to cease work and report their discovery, for, unlike in the United States, fossils in China are privileged property. "Chinese policy is you have to protect fossils," says Xiao. "Chinese know dinosaur bones cannot be made."

On the seventeenth of December 1979, as he recalls the important date in his own life, Xiao was one of several regional paleontologists called in to work on the Zigong quarry. Local administrators called in Dong Zhiming as chief dinosaur paleontologist of the Institute of Vertebrate Paleontology and Paleoanthropology in Beijing to supervise Xiao and his colleagues.

After two seasons of concerted digging, Dong and Xiao and their crews had uncovered thousands of bones, and dozens of wonderfully whole dinosaurs. But thousands more dinosaur bones likely await excavation. Although Dong would have preferred to bring the fossil bounty back to Beijing, "the [local] people," as Xiao says proudly, "refused to give up the fossils."

Most of the fossils Xiao and Dong found at Dashanpu are of the short-necked plant-eater *Shunosaurus.* It is perhaps the most common of all Chinese dinosaur fossils. At Dashanpu its skeletons lie so perfectly articulated one can count all the vertebrae, front to back. As they dug, excavators could examine the short, blunt claws of its foot, its robust spatulate teeth, and high, forward nostrils as if the fossils were already completely reconstructed. The animal's tail lay curved gracefully behind it, pulled into position shortly after death by shrinking ligaments.

At more than thirty feet long, *Shunosaurus* was small for a sauropod, with twelve or thirteen vertebrae in its neck, distinguishing it from its longer-necked cohabitants of this ancient delta. The back vertebrae provide a clue to just where *Shunosaurus* stands in the saga of sauropod evolution. The tall spines sticking up from each of these back bones are not split down the middle. This primitive characteristic distinguishes this dinosaur from the later sauropods of its own geographic neighborhood. These undivided spines link *Shunosaurus* with Early Jurassic prosauropods around the world that are thought to be ancestral to many giant sauropods.

Sauropods are the broadest and among the most enduring of dinosaur groups known. Yet despite the diversity and longevity of the group, the world's expert on sauropods, Wesleyan University professor Jack McIntosh, numbers the "fairly complete skulls" at eighteen world-

Chinese fossil workers expose the skeleton of a large sauropod dinosaur, complete with the exceedingly rare skull, from the purple mud of Dashanpu Quarry in Szechwan. [Photo courtesy New China Pictures Agency]

wide. Sauropod skulls are so rare, at least in part, because they are tiny and delicate relative to the monstrous sauropod bodies. In Dashanpu, however, dinosaur skeletons were so gently fossilized and little-disturbed that skulls are known for at least three shunosaurs, in varying stages of growth. McIntosh went to Dashanpu in 1984 to see one of the *Shunosaurus* skulls for himself. "It was so beautiful, simply gorgeous," he recalls.

The most curious of all *Shunosaurus*'s features were not found in its rare and delicate head, nor were they immediately apparent to the scientists who excavated Dashanpu. They were two pairs of short spikes and a short club on the animal's tail—reported by a Chinese researcher in 1988. While tail clubs are a well-known feature of armored dinosaurs, none had ever been found on a sauropod. To some paleontologists the club suggests that some sauropods, at least early in their evolution, used their long and whip-slender tails for defense.

Dong and Xiao discovered that *Shunosaurus,* at thirty to thirty-five feet long, shared its browsing habitat with two other varieties of sauropods, one long-necked, one short-necked. These varied in size from fifteen feet long to sixty-six feet long. Just how all these giant dinosaurs were related to each other and to the various sauropods that flourished around the world from the Middle Jurassic well into the Cretaceous era (in places until near the end of that era) is a puzzle. All the Dashanpu giants may belong to the diplodocids—long, slender, whip-tailed sauropods such as *Diplodocus* of North America, with eyes set far back on the head, nostrils right on top of the skull, and pencillike teeth at the front of a long, broad snout. So Jack McIntosh believes.

McIntosh offers his own "ninety percent conjecture" about the evolutionary role of the peculiar Middle Jurassic Chinese sauropods. "I think China may have been the cradle of evolution for diplodocids." As his best evidence, McIntosh cites the *Shunosaurus* teeth that "are halfway to being diplodocid, slender but with extended crowns. But what route the Chinese sauropods would take to get to North America [where later-Jurassic-era sauropods are best known], I'm not at all sure. They may have come to Africa and maybe they went on through Argentina to North America. We don't have any Upper Jurassic finds in Argentina, so that's all pretty iffy."

Other paleontologists, Dong among them, don't see the ancestral relationship of these giant dinosaurs to others worldwide. Rather, they believe that these Asian dinosaurs took their own evolutionary detour in the middle of dinosaur time.

What is also puzzling is how the various mysterious sauropods of the Dashanpu quarry coexisted, unless all were just passing through Dashanpu at various times in their own migrations. For all these giant vegetarians to share a habitat, no matter how luxuriant, some partitioning of resources must have occurred. Perhaps the longest-necked ones grazed only the upper reaches of the canopy, leaving lower levels to the others.

A related question of some current contention is how such long-necked sauropods could have held their heads erect. Paleontologist Peter Dodson believes sauropods could not have held their heads fully elevated without some special, and as yet unknown, anatomical adaptation such as a carotid artery capable of contracting and pumping blood through the long neck to the head.

Giraffes, with far shorter necks than the Chinese dinosaurs, have enormously high blood pressure levels. They also possess highly convoluted blood vessels at the base of their brains to help dissipate blood pressure in the region when the animal bends its head down to drink. The blood pressure needed to push blood to the head of big sauropods such as *Brachiosaurus* and *Diplodocus* has been calculated at 367 to 717 millimeters of mercury of systolic pressure, a figure unprecedented in any known living vertebrate. In comparison, giraffes' blood pressure is 240 to 280; that of healthy humans about 120 to 140. If sauropods were aquatic, as scientists once thought, the blood pressure requirements would be reduced, but as Bakker, Dodson, and others have observed, sauropods appear well-adapted for life on land.

Dodson speculates that instead of feeding in the treetops, sauropods swept their necks vacuum-cleaner style over lower levels of vegetation, a scenario also envisaged by Canadian paleontologist Dale Russell for the longest-necked creature of them all, *Mamenchisaurus*. Among other sauropods, says Smithsonian paleontologist Nick Hotton, only the brachiosaurids (relatives of the huge *Brachiosaurus*), which have a columnar neck and front legs considerably longer than back, appear to have been designed for high-level feeding. The other familiar giant sauropods of the American West, such as *Diplodocus* and *Apatosaurus* (formerly "Brontosaurus"), "carried their necks in a downward out-ward arc," according to Hotton. "If they tried to bend it to get at heights, they'd have lost half their potential height because of the curve in the neck, the way their bones are put together. So the head could not have stood up too high, no matter what."

If the huge sauropods of China and later the American West did stretch their necks high, Hotton contends, the high blood pressure needed would suggest that they were hot-blooded. "It's supposed to be true that you need endothermy [hot-bloodedness] to get high blood pressures, the systolic pressure being lower in cold-bloods."

The huge sauropods were hardly the only new and enlightening fea-ture of the Dashanpu discoveries. However the Dashanpu sauropods stood, foraged, and partitioned the plant resources, it is likely they saw small and agile herbivores tiptoeing around their feet. One such dino-saur, a juvenile, excavated at Dashanpu was named for Xiao and the quarry (*Xiaosaurus dashanpensis*). It is displayed as it was found in the quarry, nearly complete, with skull attached. In adulthood this swift plant-eater was scarcely five feet long.

The most spectacularly distinctive of all the unfamiliar Dashanpu dinosaurs was the stegosaur *Huayangosaurus*. This animal is so dis-tinct from other plate-backed dinosaurs known across the world that Xiao considers it not only a different genus, but a new evolutionary family. That is a dramatic statement by taxonomic standards, but one Bakker and Sereno (who has studied the dinosaur at length) support. Several specimens and nine skulls of *Huayangosaurus* have been turned up at Dashanpu, revealing the animal to be sixteen feet long in adulthood. With its small teeth it may have chewed on tender small leaves close to the ground that were overlooked by the bigger, stronger-toothed sauropods.

Huayangosaurus bears the signature double-row stegosaur plates upon its back. But its "plates" look more like spear-shaped projections, likely an early phase in the evolution of plate adornments. This pecu-liarity is but one of the features that suggest *Huayangosaurus* is the

East meets Wild West. Chinese stegosaur expert and curator of the Chongqing Museum of Natural History Xiao Xi Wu inspects Al-losaurus remains at the Cleveland Lloyd Dinosaur Quarry in Price, Utah, in the company of Colorado paleontologist Robert Bakker. [Photo courtesy of Robert Stahmer]

most primitive of all the stegosaurs (known from Asia, Africa, and North America). *Stegosaurus,* the familiar dinosaur from the Late Jurassic of the American West, had longer hind legs than front, but *Huayangosaurus*'s four legs are nearly equal in length, a more common anatomical arrangement.

Paul Sereno also notes that *Huayangosaurus* may have been sexually dimorphic—that is, anatomically distinct by gender: "There is a little core on one side of the top of the head of one specimen that appears in no other stegosaur. But another specimen doesn't have it. The sample size is too small for more than guesswork, but since both nearly complete huayangosaurs appear to be adult, it's possible one sex, maybe the displaying males, had horns, while females did not." The spikelike nature of the dorsal plates on all huayangosaurs suggests

Was Allosaurus *(as recreated by sculptor Stephen Czerkas) a descendant of* Gasosaurus? *[© Stephen Czerkas; Photo courtesy Sylvia Czerkas]*

these features evolved, like the clubs at the sauropods' tail-ends, as defensive weapons.

The predator against which all these herbivores sought to defend themselves at Dashanpu was the colorfully named *Gasosaurus constructus.* This killer was less than fifteen feet long and seven feet high, but equipped with strong claws and daggerlike teeth with steak-knife serrations. It is less well-preserved than its prey. Dong named it cleverly, as Xiao explains, for not only does the genus name honor the gas company that found the quarry, but " 'gas' means trouble in Chinese."

Just how this predator fit into predatory dinosaur evolution isn't clear. Dong identified *Gasosaurus* as a megalosaur. *Megalosaurus* was the first dinosaur ever scientifically described, in 1824, by British zoologist William Buckland (see Chapter 1). Buckland based his identification on a jawbone. Since then, other bits and pieces from around the world have been ascribed to the same animals, making the megalo-saurs a junk-basket category for a host of little-understood predators from mid-dinosaur time. "I wouldn't touch the megalosaurs with a fork," says Bakker of their taxonomic confusion. As for *Gasosaurus*, he thinks that "the shape of the skull seems similar to allosaurs, but it is hard to say whether there is a direct relationship."

Gasosaurus is more likely a close ancestor of another large and slightly later Chinese predator with the tongue-twisting name of *Jiang-junmiaosaurus*, excavated in China's far northwest in 1984. These big primitive hunters weren't solely found in China, though our record of them is sparse. In December of 1990 an Ohio State University geologist, David Elliot, came upon fragments of a large dinosaur near the summit of 15,000 foot high Mr. Kirkpatrick, just 400 miles from the South Pole. The creature is from the Mid- to Late Jurassic and its skull, excavated by scientists from the Fryxell Museum at Augustana College, bears a striking resemblance to *Jiangjunmiaosaurus*.

Whatever *Gasosaurus* was, it was not what killed the dinosaurs of Dashanpu, given their splendidly preserved remains. The seemingly peaceful scene of a hundred undisturbed dinosaur corpses suggested to the Chinese, at least initially, a very different reason behind the deaths.

The purplish mudstone containing the dinosaurs was composed of siltstone, and of sandstone rich in mica. This sedimentary composition characterizes a lake and river shore environment. So the dinosaurs fossilized in "a shallow bank at low energy," according to Chengdu College of Geology scientists.

But a gentle lakeside mud-packing was only part of the explanation for the ideal state of preservation of dinosaurs at Dashanpu. In a sophisticated bone analysis, Chinese scientists cross-sectioned pieces of dinosaur bones and analyzed them under electron microscopes. They found high levels of arsenic concentrated in the dinosaurs' bones. Local plants, they surmised, were peculiarly and lethally abundant in arsenic. The dinosaurs dined upon this tainted foliage and died of poisoning, a dinosaur Jonestown.

Xiao disagrees: "The dinosaurs lived around there and were not carried far by water, which is why there is no damage to the bodies. But the bones at the top of the quarry differ from the lower. They are worn smooth by water, and so are pebbles." So, Xiao concludes, the dinosaurs likely died of natural causes over long periods, and in a more conventional mode of fossil deposition, they were washed down by river currents to a collecting point for fossils and sediments. Arsenic was among the minerals that slowly entered into the bone during millennia of fossilization.

However they died, the dinosaurs of Dashanpu represent a distinctive dinosaur fauna. And they date to a time previously not known for plentiful dinosaur fossils—the Middle Jurassic. But how do we know the date? Without nearby rock of volcanic origin, chemical dating of the Dashanpu sediments was not possible. As the Dashanpu dinosaurs come from an inland basin, the traditional means for comparative dating, by the well-documented sequence of tiny marine fossils, was useless. Nor did attempts to sample the region for paleomagnetism—magnetic changes that correspond to known shifts in the earth's magnetic poles—provide a clue to the date of the Dashanpu rocks.

To date the Dashanpu fossils, a French dinosaur paleontologist named Eric Buffetaut visited museums and quarries across China in 1988. Buffetaut is energetic, fluent in English, and still in his thirties. He studied at the University of Paris, where he is a director of a large scientific research group and now researching Asian and European dinosaurs. In 1987, he married Hiyan Tong, a French-educated paleontologist who had studied geology at Chengdu University.

Buffetaut attempted to date Dashanpu by the one thing it offered in abundance—the dinosaurs themselves. As he recalls, "I compared these dinosaurs with well-dated dinosaurs from other regions. The best-dated Jurassic dinosaurs are those which have been found in marine beds in Western Europe, mainly in England and northwestern France. They are often incomplete, being the remains of carcasses washed into the sea. But they occur together with marine fossils and can thus be dated with good accuracy."

Buffetaut compared those fossils with Szechwan Basin finds. While the Chinese dinosaurs do not belong to the same groups or time as those known in Europe, he asserts, "The general level of evolution can be compared." For instance, as mentioned earlier, the Chinese stegosaur *Huayangosaurus* seems more primitive than the stegosaurs from 162 to 158 million years ago in England and Normandy.

This would seem an eminently logical process for detecting the age of dinosaur fossils. But at least two assumptions underlie such comparisons. One is that fossilized representatives of apparently primitive forms are necessarily from an earlier time than apparently more advanced dinosaurs. It is, however, possible for primitive forms to endure alongside and even outlast their more advanced counterparts.

The second, equally dangerous assumption, is that the distribution of these dinosaurs was worldwide, so that those in England can be contrasted with those from Zigong. The Zigong dinosaurs present, according to Buffetaut, "some resemblances with faunas from North America, Europe, and East Africa, which is not surprising since in the Jurassic the continents had not yet separated much."

Yet these splendid dinosaurs are peculiar to Szechwan. Perhaps these dinosaurs lived worldwide but elsewhere the climate did not produce sediment that would preserve fossils. In other regions the Middle Jurassic climate may have been one that eroded, not deposited, soil. Then no rocks, nor fossils within them, would have been made at that time.

Or perhaps the dinosaurs of Szechwan were unique to that region during the Middle Jurassic, just as their fossils are today. Mountain or water barriers in this time of high sea levels might have allowed for the evolution of peculiar Szechwan dinosaurs through their geographic isolation. What is certain is that, at present, there is nothing quite like China for Middle Jurassic dinosaurs.

The evolutionary relationship of the Dashanpu dinosaurs to the late sauropods and other giants of the American West was very much on the mind of Bob Bakker as he questioned Xiao during Xiao's 1988 tour of the great dinosaur digs of Utah. Those quarries and others in nearby Colorado and Wyoming offer evidence of a very different dinosaur community and environment at the upper boundary of the Jurassic era some fifteen to thirty million years after Dashanpu's distinctive dinosaurs thrived. And they raise intriguing questions about dinosaur metabolism, questions Bob Bakker has sought to answer in a highly controversial manner.

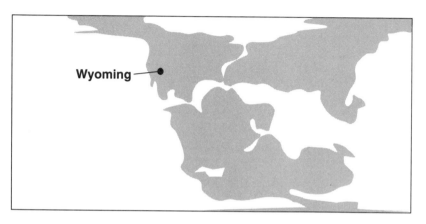

Laurasia—the northern continent—
separates from Gondwana, 163–138m. yrs. ago

T　　J　　　　C
*163-138 m.
yrs. ago*

245m.　　　　　　　　　　　65m.

AGE OF THE DINOSAURS

Triassic Jurassic Cretaceous

First animals　　　　　First land animals　　　　　　　　　First human

590m.　　　　　　　350m.　　245m. 208m.　145m.　　65m.　Today

T　　J　　　C

Heresies, Hot-Bloods, and Holes

"Bakker, is he a good scientist?"

Xiao's question came from the backseat of a rental car speeding toward Provo, Utah, and tailing a van driven by Bakker. The era of the giant sauropods, 145 million years ago, the world Bakker was showing Xiao, is the time Bakker knows best, and the alkali swamps and forests that then carpeted western North America are Bakker's research turf.

Xiao's driver slowed to answer Xiao's question. "Dr. Bob is one of the best in the world, a genius," said Bob Stahmer, Boulder photographer, journalist, and most recently Utah river-rafting guide. At the time of Xiao's question, Stahmer was also one of the latest in a long line of Bakker disciples.

Robert Bakker, more than any other scientist, is responsible for popularizing the notion of dinosaurs as swift, hot-blooded animals. Bakker attempts to answer the as-yet-unanswerable questions the public asks about dinosaurs: How did they act, look, and live? For two decades he's done so in a highly entertaining and plausible manner—casting himself and a few other researchers as heretics, tilting against an Ivy League establishment orthodoxy. Bakker's views have shaped public perceptions of dinosaurs and dinosaur science. And those views are themselves shaped by one of the most contentious careers in modern science.

Few scientists support Bakker's methods or conclusions. Yet some evolutionary biologists and paleontologists agree with Stahmer's assessment of "Dr. Bob," Stephen Jay Gould among them. The more charitable dinosaur workers acknowledge Bakker's unique talents for spinning off fresh ideas and thus stimulating research and engaging deep and widespread public interest in dinosaurs. However, few, if any, of his colleagues applaud Bakker's efforts unreservedly. And to some he is an irresponsible media hound, a slapdash researcher.

Bakker's outspokenness and willingness to speculate have attracted reporters to him. Because he holds no academic position other than unpaid adjunct curator at the University of Colorado (in Boulder), Bakker is unrestrained by academic caution. Nor can the witty, well-spoken son of an evangelist (and preacher himself in youth) resist a podium. As author, speaker, and consultant to dinosaur toy and robot makers, he earns a six-figure income while his university-employed fellow scientists eke out, at best, a precarious living. His fame and fortune explain, at least in Bakker's opinion, the profound criticisms his colleagues offer of his iconoclastic theories of dinosaur life and death.

Just how talented and telegenic a dinosaur champion Bakker is, many discovered when he spoke to his largest lecture audience to date one evening in the fall of 1988.

Underneath the huge cowboy hat he wears even at dinner, Bakker is a solid man, just a few inches shy of six feet. He looked larger, and restless and rumpled in a flannel shirt, when he stepped to the podium of a small stage at Harvard University's Yenching Library. A crowd of one hundred was present, but thousands more were listening and watching via a satellite transmission to 162 colleges and 37 corporate sites.

In rapid, booming tones Bakker tore through an often hilarious, always engaging slide lecture, illustrated with his own highly talented drawings. It's a talk he's given many times before, one he lightly labels "Hot & Cold Running Dinosaurs." He began, "Why bother with dinosaurs? Because they are Nature's special effects." He waded quickly and confidently into the "orthodox view" of dinosaurs as "stupid, slow-moving, cold-blooded . . . successful by default."

To anyone remotely familiar with dinosaurs, that view has long been supplanted by the one Bakker offers in his lecture, the one he has long touted in his talks and writings—of dinosaurs as fast-moving, nurturing, herding, above all, hot-blooded, evolutionary successes. Bakker's views on the matters of speed and metabolism, and the data Bakker uses to support them, are roundly questioned by many of his peers. To their thinking he has merely substituted one over-simplified portrait of dinosaurs for another. But in this forum Bakker marshals what is to the audience a compelling array of statistics and catch phrases.

Bakker has been proselytizing for warm-blooded dinosaurs for two decades. He was a brilliant student of many subjects, including the dinosaurs he first became enraptured by in a *Life* magazine pictorial he encountered at the age of ten. At nineteen, Bakker first prospected for dinosaurs in the West. He wrote his first scientific paper as a Yale undergraduate, arguing that brontosaurs (small *b* to include *Apatosaurus* and its close sauropod kin) were terrestrial browsers, not swamp-bound sluggards, as many scientists still argued. He spent a year and a half as an undergraduate on a "Scholar of the House" project of his own devising—alignment of the forelimbs of dinosaurs. From his knowledge of reptilian evolution, he suspected that dinosaurs walked with more erect front limbs than in many fossil reconstructions of the day, which were structured according to prevailing notions of dinosaurs as slow-going reptiles. As he has done many times since, Bakker used highly creative means to demonstrate his theories. And as is often his goal, Bakker sought to overturn what he perceived as rigid scientific orthodoxy supported by twisted a priori reasoning he calls "pretzel logic."

Bakker says of his early work, "What struck me in the sixties was that if you turned dinosaurs upside down . . . if instead of being cold-blooded and stupid and slow in the swamps, they were warm-blooded and smart and lived on the land, it started making a lot more sense."

Bakker bought alligators and lizards, which he filmed with a hand-held camera as they crawled about a Yale museum basement. Bakker reasoned that since alligators were large reptiles that share a common ancestor with dinosaurs, the gators' posture would indicate something of how dinosaurs carried their front limbs. Indeed, he concluded alligators sprawled far less at the elbow than did the museum specimens of horned dinosaurs. (He also found via treadmill energetics studies on crocodiles that "sprawlers" were more, not less, energy efficient than upright walkers such as dinosaurs, overturning previous theory.)

Bakker worked with Yale paleontologist John Ostrom, his primary undergraduate adviser, and pursued with keen interest Ostrom's developing ideas of a dinosaurian ancestry of birds, then a highly unusual view.

At Harvard University graduate school in vertebrate paleontology, Bakker followed these self-described "heresies" with many more provocative departures from accepted theory. Stephen Jay Gould, a man not liberal with his compliments, had Bakker as a student and found him to be "one of the most interesting people I've ever known. I used to keep a book recording whatever good thoughts came my way. Bakker's name was always in it."

While still a graduate student at Harvard, in 1975, Bakker published his most controversial article, "Dinosaur Renaissance," in *Scientific American,* elaborating his views of hot-blooded dinosaurs and their close relationship to birds. For the "Dinosaur Renaissance" article Bakker drew upon his own precocious scientific publications on dinosaur locomotion, behavior, and energetics. And he drew on Ostrom's discovery of an apparently lithe and highly active predator, *Deinonychus,* and Ostrom's arguments for *Archaeopteryx* as a bird that evolved from swift-running dinosaurs such as *Deinonychus*'s theropod ancestors.

Following Harvard, Bakker won a prestigious junior faculty position as assistant professor of geology at Johns Hopkins. There Bakker continued to turn out creative and influential work. He wrote on patterns of extinction in the fossil record, discussing the effects of environmental stress on dinosaur species and noting that there were fewer and fewer kinds of large dinosaurs near the extinction. In a paper for the prestigious science journal *Nature,* Bakker proposed that the evolution of modern flowering plants and herbivorous dinosaurs provided an important example of coevolution. He reasoned that the evolution of big-beaked, close-to-the-ground foraging dinosaurs more than 100 million years ago, replacing the dominant stegosaurs and giant sauropods, opened the way for the spread of flowering plants. This clever and controversial hypothesis was faulted by paleobotanists and paleontologists alike for the slim supporting evidence, yet heralded for its originality.

Bakker has continued publishing new theories and species identifications (in the popular press and journal he founded) that, if less controversial than his early work, have nonetheless met with widespread public attention and professional criticism.

He has already received abundant press attention in advance of any publication in refereed scientific publications for finds from his ongoing excavations at Como Bluff, Wyoming. Como Bluff is a site of dinosaur discoveries since railroad workers found sauropod bones there in 1877. Bakker's recently headlined finds include bizarre tiny mammals and turtles, and what he calls the oldest brontosaur and the largest Jurassic carnivore. He dubs all with highly imaginative names, such as a Cretaceous predator he planned in 1991 to name in honor of Arnold Schwarzenegger.

Also in 1991, Bakker shared with *Discover* magazine his attempt to apply his theory that continental connections fostered the spread of extinction-causing epidemics in dinosaurs. Bakker sought to document such an extinction at the Jurassic-Cretaceous boundary 145 million years ago via his discoveries of only small (and so, he thinks, more disease-resistant) dinosaurs at Como Bluff. Fellow dinosaur paleontologists were not convinced. "This is a leap off the cliff of knowledge into the abyss of speculation," says University of Pennsylvania paleontologist Peter Dodson.

Bakker summarized much of his own evolutionary research and marshaled his arguments for dinosaur warm-bloodedness in his 1986 book, *Dinosaur Heresies,* unaffectionately dubbed "Dinosaur Hearsay" by critical fellow paleontologists. Nearly five hundred pages long, it is replete with intriguing theories and Bakker's own able illustrations of nimble, leaping dinosaurs. Peter Dodson, reviewing the book in *American Scientist,* was troubled by Bakker's maverick stance. "Whereas other scientists see farther because they stand on the shoulders of giants, he sees farther because he stands on the slain bodies of his opponents. His straw man is the 'muddleheaded, orthodox paleontologist' who saw dinosaurs as sluggish, dim-witted, swamp-slogging behemoths." Bakker was lancing at a corpse that had been buried fifteen years before, wrote Dodson. "For instance, orthodox paleontologists do not today maintain that sauropods were aquatic—he himself laid that to rest in 1971. It is self-serving to pretend that shorthaired, tie-wearing orthodox paleontologists still believe that." As another paleontologist, Hans-Dieter Sues, succinctly put it: "Bakker's is the worst kind of science—pushing titans off their clay feet."

While the most extensive professional criticisms of Bakker's findings were launched by a former protégé, Denver Museum of Natural History curator Kenneth Carpenter, the breadth and depth of the objections raised to Bakker's theories are difficult to dismiss on personal grounds. For one example, Wellesley College paleontologist Emily Griffin wrote that Bakker played "fast and loose with anatomical detail," underestimating *Stegosaurus*'s brain size, falsely linking the cerebrum's function to hearing, and neglecting to discuss the relationship between high activity and brain size while concentrating on unusually large-brained small dinosaur predators.

Long on theory, short and sloppy on data; these are the charges that fly most freely concerning Bakker's work. His professional critics believe his research would be improved by the substantiation necessary to gain approval in independent, peer-reviewed scientific journals. But for many years Bakker has steered away from these publications, troubled by the restrictions of length and the possible personal animosity of the anonymous expert reviewers assigned to submissions. His principal vehicle for scientific writing has been, since 1988, *Hunteria,* a University of Colorado publication he founded. *Hunteria* submissions are assigned to reviewers before publication, but to many colleagues, the oversight isn't sufficiently rigorous to ensure careful science. "At

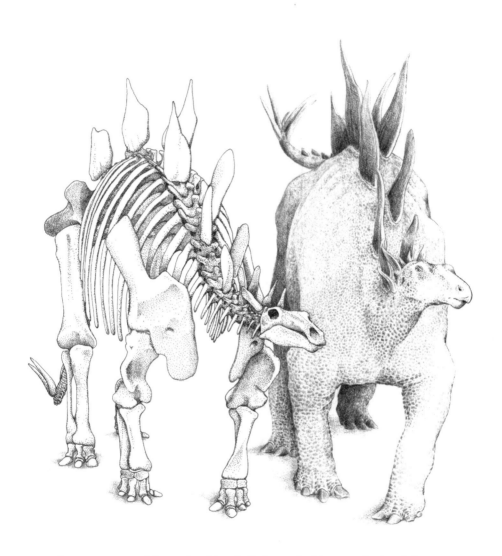

Stegosaurus, *with and without flesh and skin, as depicted by Donna Braginetz. Artists differ on whether* Stegosaurus *had alternating paired rows of plates or a single row. Plate arrangement in* Stegosaurus's *ancestry (see* Huayangosaurus*) suggests paired plates. [Donna Braginetz]*

Stegosaurus lived in the same environments as the largest of all dinosaurs, the giant North American sauropods. Here Stegosaurus is depicted with a single row of plates by sculptor Stephen Czerkas, after extensive investigation of Stegosaurus remains. [© Stephen Czerkas; photo by Sylvia Czerkas]

best it's an in-house journal, similar to Cope and Marsh creating their own journals a century ago," says Peter Dodson.

But it is the central thesis of Bakker's work—dinosaur hot-bloodedness—upon which Bakker has built his shaky scientific reputation.

Dinosaurs: Hot, Cold, or Neither?

Were dinosaurs warm-blooded? In the public consciousness, the question appears to have been largely resolved in the affirmative. But according to most paleontologists who've examined the issue, Bakker excluded, the answer to the question is: "Some," "Sometimes," or in the Talmudic tradition of a question for a question: "Who knows?"

Scientifically, the issue of dinosaur warm-bloodedness may be "a red herring" as Dodson wrote in 1981, "there now being nearly universal agreement that dinosaurs, by virtue of size alone, had relatively constant elevated body temperatures." For contemporary scientists the dinosaur metabolism issue is not whether *some* dinosaurs were warm but how they got warm and how they stayed that way.

Before Bakker began addressing the question of dinosaur warm-bloodedness in the late 1960s, the accepted opinion was that dinosaurs were cold-blooded. The work of a few researchers, and Bakker's publications, prompted a reexamination.

In 1969, Ostrom published his description of his 1964 Montana find, *Deinonychus*. That small predator's slender legs and huge toe claws suggested a fast-running, high-kicking lifestyle not associated with any cold-blooded reptile. Reptiles can run fast, but only for short distances. Then they must stop to pant, like an exhausted sprinter.

Armand de Ricqlès, a French scientist specializing in the structure of fossil bone, sought to answer the question of dinosaur warm-bloodedness empirically in the 1960s and 1970s through laboratory studies of dinosaur bones. He made thin sections of hundreds of bones of various fossil animals. Under the microscope he found evidence that Bakker in his book called "overwhelming and incontrovertible" of "warm-blooded growth patterns in dinosaurs."

What Ricqlès found was bone composed of collagen fibers connected in an irregular manner, rather than in straight rows. This pattern indicated to him that the bone was formed in a rapid growth spurt. Bakker contends that cold-blooded reptiles do not grow this way in the wild. Rather, the bones of cold-blooded reptiles exhibit growth rings, more regular growth, and fewer canals.

Today, reptiles do grow slower in the wild than do mammals, but their growth rates are highly variable. In breeding farms, alligators grow to four feet—half of adult size—in a year. Ricqlès has cited studies in which young, farm-grown crocodiles showed the fast-growing pattern of woven bone. A detailed bone study made in 1987 found dinosaur bone a combination of reptile and mammal bone patterns. The dinosaurian bone structures Bakker cited as uniquely warm-blooded turn up in many large, long-lived animals, both cold- and hot-blooded.

And contrary to Bakker's assertions, Ricqlès himself has doubts about dinosaur hot-bloodedness. He now says that "at least some dinosaurs were able to sustain growth rates as high as large mammals today, and . . . they could be long-lived, too. But what that means exactly in terms of physiology is of course still an open question. The issue has been spoiled by journalistic 'overheating' and psychological adherence of scientists and laymen to a strictly dichotomous view of the issue."

Bakker has added other lines of evidence to support his warm-blooded dinosaur scenario. He began arguing scientifically in the early 1970s that dinosaurs could run rapidly. He based his conclusions on his own interpretations of their stance. Dinosaurs' limbs moved directly beneath their bodies, not to the side, as in lizards and crocodiles. This was an adaptation for speed, Bakker contends.

But at least some dinosaurs may not have walked upright after all. In 1990, Ostrom and Rolf Johnson of the Milwaukee Public Museum tested upright and sprawling postures on a flexible fiberglass-cast model of the shoulder and arms of *Torosaurus,* a huge horned dinosaur akin to *Triceratops.* In the upright position, the *Torosaurus* elbow joint was thrust into an untenable position. The scientists suggest that all horned dinosaurs may have maintained a sprawling or semi-upright, lizardly forelimb posture. They could not, therefore, gallop like a thirty-mile-per-hour white rhino, as Bakker has suggested.

Moving fast does not in itself argue for warm-bloodedness, but Bakker reasoned that sustained high speeds suggested rapid foraging

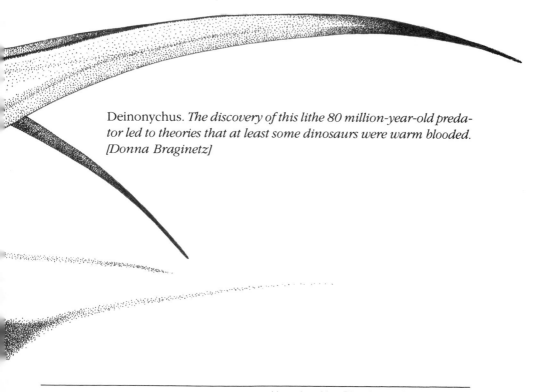

Deinonychus. *The discovery of this lithe 80 million-year-old predator led to theories that at least some dinosaurs were warm blooded.* [Donna Braginetz]

rates for dinosaur predators. They hunted quickly because as warm-blooded animals they had high food requirements. From dinosaur footprints, Bakker says he has calculated a sustained pace of four miles per hour for a ten-thousand-pound *Tyrannosaurus rex.* That is a "hot-blooded cruise" to Bakker, a speed he reckoned no known cold-blooded animal could maintain.

But no one has ever found a footprint of *T. rex,* nor has any large-predator footprint assemblage yielded a reliable estimate of four miles per hour for "cruising" dinosaurs.

Another of Bakker's innovative arguments for dinosaur warm-bloodedness was that warm-blooded dinosaur hunters, with their high caloric needs, would have needed more prey animals to munch on than would cold-blooded killers. So by measuring the ratio of dinosaur predators to dinosaur prey, Bakker thought he might find evidence of dinosaur warm-bloodedness. And indeed, he found predator-to-prey ratios to confirm his theory. But other researchers have found very different ratios, and many believe the fossil record is so uneven, due to variation in the way fossils are preserved and collected, that all predator-prey ratios are unreliable.

Cautious dinosaur researcher James Farlow of Purdue University goes further, pointing out a key weakness in Bakker's method—a failure to compute the rate of turnover in the prey population. Small victims, the rabbits of their day, reached maturity sooner, and so replaced their numbers faster. And small predators would have lower food requirements than large ones. In 1976, Farlow, then a believer in dinosaur hot-bloodedness, attempted to apply the method Bakker used to a collection of some of the best-known dinosaurs, from the Judith River Formation of Alberta, Canada (see Chapter 1). His results confounded him. "If those dinosaurs were hot-blooded, they would have starved to death."

Bakker suggested another proof of warm-bloodedness for the vegetarian dinosaurs. Hot-blooded huge sauropods with small mouths would have had to ingest huge quantities of vegetation to stay warm and moving. To help them do so, their stomachs acted as huge vats. There the work of digestion was carried out, abetted by the grinding effect of gizzard stones. Most scientists now back this feeding strategy for the largest dinosaurs, but few think that in a seasonally dry land, dotted with conifers and ginkgo trees, a herd of giant animals could have found enough food to maintain the high energy demands of a hot-blooded lifestyle. Unless they were feeding on some unknown plant extraordinarily high in protein, John Ostrom has concluded, these dinosaurs would have had to eat more than a hundred pounds of plants a day and processed it quickly, something they show no specializations for doing.

But the giant dinosaurs wouldn't have needed to burn vast amounts of fuel to stay warm. Their own mass would warm them. Like elephants, they would derive body heat from the activity of their internal organs. Since they were so large, they would have cooled off slowly once warm. Rather, as Dodson suggests, "they probably had to stay out of the sun in midday to keep from suffering meltdown."

Bakker also suggests that dinosaurs lived a life that only befits hot-bloods, "in the fast lane." He views a whole range of dinosaur behav-

Robert Bakker inspects a cast of the jaw of Allosaurus *at the University of Colorado. [Photo by author]*

iors, from head-butting to hooting and snorting, as facilitated by evolutionary modifications for "high-energy aggression." Other aspects of dinosaur anatomy also argue for warm-bloodedness, according to Bakker—their well-developed vascular systems, for instance. Bakker's additional evidence includes the wide distribution of dinosaurs, even into arctic latitudes, and the fast rate of dinosaur speciation (an assertion not well documented).

"There is no middle ground for him—no lukewarm dinosaurs, no cold sauropods," says Dodson. It is this "either-or" reasoning, as much as questions about the data and conclusions of specific aspects of Bakker's arguments, that most trouble other paleontologists. Animals do not simply run hot or cold; their metabolisms are considerably more varied (as paleontologist Jack Horner makes clear in his and James Gorman's book, *Digging Dinosaurs*). Endotherms (meaning "inside heat"), or warm-blooded animals, regulate their body temperatures at a constant level. Ectotherms ("outside heat") also regulate their own temperatures (by moving into the sun, for example), but their body temperature responds directly to outside temperature. That's as far as most of us, and Bakker's book, go.

But temperature regulation is more complex than this. Some animals maintain a relatively high and constant body temperature. These are homeotherms. Larger animals can use their own bulk to maintain high body temperature without high metabolic output. They can do so because surface area is inversely proportional to volume: the bigger the animal, the smaller as a percentage is its surface or skin. Heat is lost in proportion to surface area. Mice are little and light and lose heat rapidly, so they need a fast metabolism to stay warm and active. Elephants, also endotherms, don't have to work so hard to stay warm because they retain heat better, benefiting from mass homeothermy.

On the other hand, ectotherms, or "cold-bloods," undergo a wide variation in temperatures according to their environment.

The methods used to achieve temperature regulation also vary. Most, but not all, endotherms have a high metabolism, all the time. Most ectotherms have a slow resting metabolism. But among the warm-blooded mammals, bats and hummingbirds are mammals that lower their metabolism at night or when food is scarce. What metabolism

have they? On the supposedly cold-blooded reptilian side, large leatherback turtles can keep their body core temperature eighteen degrees
centigrade above water temperature as they migrate six thousand miles
per year through cold arctic waters. Are they warm- or cold-blooded?

Applying their studies of the leatherback turtles to dinosaurs in 1991,
James Spotila of Drexel University, Dodson, and colleagues at Purdue
University concluded that it didn't take much energy to keep a dinosaur comfortable. Even very small and uninsulated dinosaurs would
have had body temperatures close to that of their environment whether
hot- or cold-blooded. The scientists suggest a metabolic model for
dinosaurs called *gigantothermy,* the system whereby large body size,
low metabolic rate, insulation, and controlled circulation help large
living reptiles maintain constant and high body temperatures without
the energetic expense of being truly warm-blooded.

Where does all this leave us? The hard evidence for hot-blooded
dinosaurs is still lacking. But Jack Horner thinks he may have it—and
once again it lies in the structure of dinosaur bone.

Having discovered an unprecedented volume and age range of
duckbilled and other dinosaur fossils in Montana (see Chapter 10),
Horner set about addressing the question of dinosaur warm-bloodedness. Horner's most celebrated discoveries were the nests of the
twenty-five-foot-long duckbilled dinosaur he named *Maiasaura*
("good-mother reptile"). Within the nests Horner found the tiny bones
of young dinosaurs fourteen inches to three and one-half feet long.

The disparity in nestling sizes suggested to Horner that these dinosaurs were warm-blooded. Why? "It takes a bird a month or two to
double in size," he and Gorman wrote. "But it takes an alligator, a fairly
sophisticated reptile, a year. No known creature stays in the nest that
long. So if these animals did stay in the nest, as we concluded (from
fossilized regurgitated food, trampled shells, and surrounding nest
materials), and if they grew at the rate of cold-blooded animals, there
was an insoluble problem."

Looking for evidence in the bone structure of these fast-growing
young duckbills, Horner and a precocious young protégé, Jill Peterson,
cut razor-thin sections of dinosaur bone and examined them under a
microscope on the admittedly vague notion that looking at young

duckbill bone would reveal whether the animals were warm-blooded and fast-growing. But the project soon grew to be much more, an ongoing study that offers promise of some of the few current experimental insights into the issue of dinosaur hot-bloodedness.

Peterson, a tiny, blond, and blue-eyed young woman, began corresponding with Horner over a decade ago. She was then a fifteen-year-old in a New Jersey high school so overcrowded that its administrators urged students to pursue projects in the community during schooltime. Horner was a fossil preparator at nearby Princeton University. Peterson had read of his nascent Montana dinosaur research in the *New York Times*. The next summer was the first of several she spent in the field with Horner.

After her high school graduation, in 1982, Peterson managed and codirected Horner's field work at Montana State University. The next year she was appointed a research associate of the university and began her bone-growth studies with Horner.

In 1985, while still an undergraduate, she presented the results of her bone study at the annual SVP meeting. Her objective was not to address dinosaur warm-bloodedness as much as to judge the animals' ages—to discern patterns of bone growth corresponding to different stages of dinosaur development. Others had tried before, dating back to the pioneering work of the eccentric Transylvanian paleontologist Baron Von Nopcsa, who distinguished juvenile duckbilled dinosaurs from adults by their rib-bone structure. But no one, before Peterson, had succeeded in doing anything more with bone identification than discerning adult dinosaurs from juveniles.

What Peterson found were six separate growth phases in *Maiasaura,* characterized by different stages in development of structures known as osteons, microscopic layers of bone around a central canal. Newly deposited as bone tissue in babies, osteons become eroded and redeposited as the animal matures. "It was excellent work and professionally presented," recalls Bakker.

Peterson's elegant study won her the Romer Prize for best student lecture, over the research of doctoral and postdoctoral students many years her senior. Her study said much about rapid development in dinosaurs, but it did not resolve the issue of duckbilled dinosaur warm-bloodedness. As Ricqlès pointed out to Horner, rapid growth in dinosaurs might be related to climate, not endothermy: "What if the environment was hot enough that cold-blooded animals grew like a house afire?" Further research was called for.

Peterson's research paper marked the end, not the beginning, of her career and her relationship with Horner. She is now Jill Rife, a middle-school librarian, married to high school science teacher and living in a tiny Colorado mountain town. Horner, however, has proceeded with the work they began together. Ricqlès suggested that a growth series from contemporary endothermic and ectothermic animals, living in the same environment, would be a necessary baseline for comparing dinosaurs and other extinct animals, hot- and cold-blooded. So in cooperation with the veterinary science school at Montana State University, Horner raised an emu, a cow, and a crocodilian known as a caiman, "two hot-bloods and a cold-blood," implanting platinum wire bands around their femurs at various stages of growth. Once they reached adulthood they were killed and dissected. Horner is now examining and comparing the different growth patterns.

Horner hasn't drawn his conclusions yet, but he writes in *Digging Dinosaurs,* "I suspect that mass homeothermy was important, but not as a complete explanation. Young dinosaurs were too small for that strategy to hold." As he told *Newsweek* in mid-1991, Horner thinks dinosaurs "were specialized to grow real fast. But after they reached sexual maturity, or a particular size, their metabolism slowed way down so they didn't require so much to eat."

Paleontologist James Farlow of Purdue University, who has researched dinosaur metabolism as much as any paleontologist, offered his own intriguing speculations on dinosaur metabolism in 1990. Perhaps the growth rings Ricqlès saw in dinosaur bones were seasonal markers, suggesting to Farlow that dinosaurs varied their metabolic rates by age, even by season. Thus, dinosaurs may have been individually hot-blooded *and* cold-blooded. In which case, Farlow concludes, "perhaps we will have to start calling them damn good reptiles."

Dinosaur metabolism will likely remain an unanswerable mystery. "Short of sticking a thermometer into a dinosaur," says Farlow, "I don't see how we'll ever know."

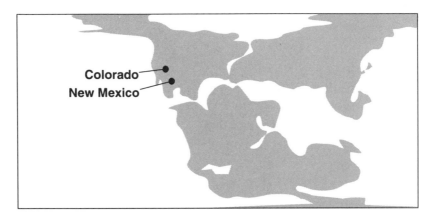

Laurasia and Gondwana, 163–138m. yrs. ago

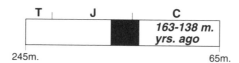

T   J   C

163-138 m. yrs. ago

245m.   65m.

AGE OF THE DINOSAURS

Triassic Jurassic Cretaceous

First animals   First land animals   First human

T   J   C

590m.   350m.   245m. 208m.   145m.   65m.   Today

Land of Giants

One hundred and forty-five million years ago the American West rumbled with the footfalls of the largest animals ever to walk the earth. These were creatures named, justifiably, "thunder lizard," "super lizard," "ultralizard," and "earth-shaker lizard"—titans as long as one and one-half blue whales, and three times as tall as giraffes.

Bearing in mind that 80 percent of all living animals are smaller than even the chicken-sized dinosaurs, and the half of all dinosaurs weighed less than two tons, the giant sauropod dinosaurs of the Late Jurassic were stupendously large by any standards, even dinosaurs'.

The bones of the giant sauropods are scattered throughout the world, and best known from the American West. The most spectacular and celebrated of these titanic remains belong to the biggest animal that ever lived. Finding such a record-setting speci-

Sauropod trackways, Glen Rose, Texas. These famous tracks, the first ever known of a sauropod dinosaur, show the stride of some of the giant North American sauropods. [Photo courtesy James O. Farlow]

men is not a scientific problem of great concern to paleontologists. But the biggest-dinosaur hunt is a contentious quest that has occupied the energies of prominent researchers and the attention of the public over the last several years. Much more is at stake in this tape-measure battle than some bragging rights—the future of productive research projects and the legacy of one of the last of the nonacademic dinosaur diggers, "Dinosaur Jim" Jensen.

From the age of ten, on a farm in the Utah mountains, Jim Jensen imagined himself digging dinosaurs. As he told John Noble Wilford of the *New York Times* in 1981, "While some boys dreamed of a new bicycle, I dreamed of finding dinosaurs. I would always wake up before I could dig them up."

But Jensen did get his dinosaurs, though his scientific education fell short of completing high school. He worked in Alaska as a mechanic, then in Massachusetts as a jack-of-all-trades for Harvard University mammal paleontologist Alfred Romer (see Chapter 2). With Romer, he looked for dinosaurs in Antarctica and found them in Argentina. Jensen left Harvard on a temporary assignment in 1961 to construct an allosaur mount for Brigham Young University. While he was there, a nonteaching position opened in the geology department and Jensen stayed until he was unwillingly retired in 1983. In that time, Jensen collected his way across Utah and western Colorado on a shoestring budget, twenty-six quarries in twenty-three years, digging from late April to early September each year. He dug thousands of dinosaur bones, far more than can be prepared to this day. What Jensen did excavate, prepare, and name—the two biggest dinosaurs ever found—made him and his institution famous.

Now in his seventies, a big, rawboned, voluble man, Jensen has turned away from paleontology, but reflects with happiness and regret on his role in locating and naming the biggest of dinosaurs.

Though famed for his talent for finding dinosaurs, Jensen actually discovered few, if any, of his excavation sites. "I'd check out reports from people who knew where to take things," Jensen acknowledges. Dry Mesa, in far western Colorado, was one such site, a high hillside site above the convergence of two deep canyons, their slopes studded with piñon and juniper trees. Sawmill operator and amateur collector

Seismosaurus, Ultrasaurus, and Supersaurus (and artist) [John Sibbick]

Eddie Jones took Jensen to the quarry after Jensen had admired a huge bone of a Dry Mesa carnivore on Jones's coffee table.

In the summer of 1972, Jensen went to Dry Mesa on a headline-making hunt. "I was an obsessed man. I was going to find the oldest, biggest carnosaur [meat-eating dinosaur], the bigger the better, even though scientifically that means little."

Late in the summer, still in the exploration phase of quarry activities, Jensen drove a bulldozer across part of the site, making a three hundred-foot-long cut. As his longtime assistant Ken Stadtman recalls, "All along that cut, there was bone everywhere we looked."

Jim's son, Ron, was working by himself on a section of the cut on a long-dried streambed, some hundred yards from camp. He struck upon the shoulder blade of an enormous dinosaur, but continued working, silently, until he had uncovered it completely. Then he hid it under burlap bags. "Ron said, 'Get your big butt over here,' " Jim Jensen remembers. "He whipped off the bags and there it was, eight feet long! It blew my mind."

A little figuring, based on comparisons of this shoulder bone and neck vertebrae found nearby with those of other giant sauropods, put the length of the new animal at eighty-two to ninety-eight feet (or even one hundred forty feet if it proved to be related to the long-tailed *Diplodocus,* as Jensen believes) and its height at fifty-four feet—a new "biggest dinosaur," far larger than the then record-holder, the stout, eighty-two-foot-long *Brachiosaurus.*

In 1979, Jensen was again digging around the Dry Mesa site, under the gaze of a Japanese television crew, when he and his crew came upon another shoulder blade. This one was nearly nine feet long, and broader than that of *Supersaurus.* If not from an animal longer than *Supersaurus,* it certainly belonged to a far heavier animal.

But a new dinosaur isn't officially recognized until it's formally named, usually in a scientific paper that gives precise descriptions and measurements of bones and compelling evidence from that data for concluding the bones could not have belonged to a known animal but had to have belonged to one of an unfamiliar age, size, or sex.

Particular care is taken with such descriptive papers, for according to the arcane rules of zoological nomenclature, wherever a find is referred

In retirement outside his home in Provo, Utah, "Dinosaur Jim" Jensen still dresses like a field man. [Photo by author]

to first by any name, in any publication, that name is permanent. So is the accompanying description, however partial or misleading. If the name turns out to be previously claimed, or the description invalid, the new species claim is dismissed and the entire naming process must begin again. More than mere ritual or formality, this is a universally agreed-upon code of conduct for a science, a vital means of quantifying and verifying the inventory upon which all paleontological business is conducted.

Jensen is not a trained scientist, nor is he temperamentally inclined to bookish research and scholarly writing. He was quick to name his finds but slow to produce scientific papers doing so properly, perhaps from unfamiliarity with the format, other commitments, or problems with documentation—he did not make thorough quarry maps or write thorough diaries at the time of discovery. Nonetheless, he did write up his finds, seeking to establish his claims for the biggest of all dinosaurs.

Jensen did not get around to formally dubbing *Supersaurus* and *Ultrasaurus* until October 1985, after his retirement. He did so in a rather obscure Brigham Young University journal, *The Great Basin Naturalist,* and in a manner that left his assertions on unsteady ground.

Why? Unknown to Jensen, by 1985 the *Ultrasaurus* name had already been scientifically claimed, by a South Korean paleontologist, Haang Mook Kim. Kim wrote a series of short papers describing South Korean dinosaurs including bits of a dinosaur he thought larger than *Supersaurus,* dubbing it *Ultrasaurus tabriensis.* According to colleagues, Kim's find turned out to have been misidentified, the fossils poor in quality, and the dinosaur considerably smaller than Kim believed.

But there were other problems with Jensen's naming of his giants. Though based on prepossessing fossils, those fossils are but fragments from a jumbled quarry. And they could only be contrasted with other fragments of already named but still only partially known dinosaurs. And Jensen made some strange choices among the bones he'd found as the basis for identifying his new giants.

Based on the long spool on the backbone, Jensen described *Ultrasaurus* in 1985 as one of the brachiosaurs, solidly built, giraffe-necked sauropods with front legs longer than back legs. Jensen didn't

Jim Jensen reconstructs the forelimb of the tallest and heaviest known dinosaur, an animal he excavated, described, and named Ultrasaurus. *[Photo by Mark A. Philbrick, Brigham Young University]*

base his new genus on the intriguing giant shoulder blade evidence. Rather he used one of several vertebrae found nearby as his holotype—the distinctive bone upon which hangs new species or genus identification. He didn't assign *Supersaurus* to any family, though its ribs seemed to Jensen to be like the relatively slender, whip-tailed *Diplodocus* and its kin. He included no diagnosis at all with his description of *Supersaurus*.

Jensen called his own diagnostic skills into question in a 1987 paper. There he took the unusual step of withdrawing his assignment of a neck vertebra to *Ultrasaurus*. He redubbed it diplodocid, in keeping with the backbone's forked spines.

And Jensen noted, the biggest scapula from Dry Mesa should be the holotype for identifying *Ultrasaurus*. Now, however, he has changed his mind again: "That was wrong. I was pressured to publish a lot quickly. I assumed incorrectly it was *Ultrasaurus*." Now Jensen believes the shoulder blade belongs to an unidentified brachiosaur, but not *Ultrasaurus*. By this latest analysis the quarry at Dry Mesa would contain at least four different giant sauropods.

Not surprisingly, Jensen's reconsiderations of *Ultrasaurus* left other paleontologists puzzled and displeased. Several doubt that *Ultrasaurus* is an animal distinct from *Brachiosaurus,* and few think Jensen's description and naming will stand the test of time. But the Dry Mesa quarry is still producing new fuel for the dispute.

Early on the morning of August 18, 1988, Brigham Young University preparators Cliff Miles and Brian Versey began their day's work at the Dry Mesa quarry. The morning was crisp, the view dramatic; but from that view, Miles and Versey were looking away—toward the sandstone and clay of the hillside. Under the supervision of Jensen's former assistant, now the university's paleontological field director, Ken Stadtman, they were continuing a bit of excavation begun the month before. Then Miles's father, visiting the site, found a protruding dinosaur hip bone.

Two hours later, Miles asked Versey to pick away at the wall of sandstone to give him more room to get around the fossilized bone. Within minutes, Versey heard the distinctive dull sound of pick striking fossil and called, "Cliff, I found bone." Digging several feet away, Miles also hit upon fossil. Hours later, following the line of bone, they met in the middle and realized they were excavating a single bone. When, with a few days of strenuous labor, the surrounding stone had been cleared away, it was apparent Versey and Miles had come upon an immense bone. It was, in fact, the largest bone complex ever found— the pelvis of a huge sauropod dinosaur. Once cleaned, the pelvis measured six feet in length and 1,500 pounds in weight.

Kenneth Stadtman of Brigham Young University and co-worker excavating the largest single dinosaur bone ever found, the six-foot-wide pelvis of what they believe to be a Supersaurus. *[Photo by Mark A. Philbrick, Brigham Young University]*

Whether or not this bone belonged to the largest dinosaur that ever lived was another matter, one of considerable public interest and professional dispute. From the pelvis's location at the site, twenty yards from those specimens previously referred to as *Supersaurus* and *Ultrasaurus,* there was no telling which dinosaur had left this fossil. After comparing it to other sauropod hips, Stadtman thinks it likely belonged to *Supersaurus.* But the day after the find was announced to the media, Jensen himself was on the scene, for the first time since his abrupt retirement from the university five years before. He declared the pelvis to be from the more robust *Ultrasaurus,* for it shows a tall spine, much like that on the vertebra he originally used as the animal's holotype and which, he thought, came from the animal's back, near the pelvis.

David Gillette, Utah's State Paleontologist, has overseen the excavation, preparation, and study of the largest dinosaur ever found, New Mexico's Seismosaurus. [Photo by Beth and Greg Henry]

One interested party who didn't join in the dig or the dispute was the newly appointed Utah State Paleontologist David Gillette. Gillette was more than busy coordinating his own dig, in New Mexico, begun in 1985 and ongoing as of 1991, for what he is convinced is the world's longest dinosaur. It's a long-tailed sauropod he calls *Seismosaurus* (named for the ground-shaking effect it must have produced when walking), an animal of a staggering length estimated at from 110 feet to 150 feet long—as long as half a football field.

Some of the huge and half-cleaned bones of *Seismosaurus* cram Dave Gillette's office—two small rooms, jammed with fossils in various states of preparation, in the basement of a converted train station in Salt Lake City. Gillette is a small man with an outsized sense of humor, a flaring temper, and far-ranging curiosity and enterprise. Happily for him, the state paleontologist's job often takes him out to field sites, museums, and courtrooms (to help prosecute illegal fossil collectors; see "Selling Fossils," page 195) and away from the tedious business of cleaning *Seismosaurus*. On weekends he's often away in New Mexico, where he and his helpers have exposed most of what they think they can find of the world's longest dinosaur.

Like Jim Jensen, Gillette has become a paleontological celebrity, identified with the biggest of the big. And like Jensen, he's been slow to publish his results, casting a cloud of uncertainty over his findings. But if Jensen sought dinosaur greatness, Dave Gillette would say greatness, at least a great dinosaur, was thrust upon him.

Gillette never set out to work on dinosaurs. He earned his Ph.D. at Southern Methodist University, writing on giant armored mammals. Gillette eventually returned to SMU to fill his retired adviser's chair in the geology department. Gillette was peripatetic as ever, making working trips abroad, to France, Japan, the Libyan Sahara, Baja California, India, Anatolia, and Argentina (he visited Ischigualasto but wasn't allowed to excavate there).

From SMU, Gillette went to the New Mexico Museum of Natural History in Albuquerque in 1983. There he worked as paleontology curator before taking the Utah post in 1988, and it was there he first learned of *Seismosaurus*. The dinosaur was, in fact, found several years before he came to New Mexico. A group of backpackers, toting a video camera, happened upon several large bones, amid Indian petroglyphs, protruding from the crumbling side of a mesa on Bureau of Land Management terrain. It is a handsome site, as Gillette describes it, "a long drive down a jeep trail through rugged country at a canyonhead. It's sagebrush and juniper habitat in quite colorful desert, mainly reds and greens."

Gillette petitioned the BLM for a permit to excavate the site, permission he says he was able to obtain for "only three weeks over the next eighteen months," thus delaying his research. But in one frenzied weekend, Gillette and ten others excavated eight tail vertebrae, still aligned. The fossils, locked in a matrix of sandstone virtually indistinguishable from the bone, proved extremely difficult to remove. The team quit only upon reaching a point where the fossil "moved into the hill," as Gillette puts it.

High-Tech Dinosaur Digging

The same year, 1985, Gillette began work on *Seismosaurus,* he began turning his excavation into what has become one of the largest and most technologically enterprising of paleontological studies.

The project began serendipitously and without great eagerness on Gillette's part. As "a nineteenth-century paleontologist," he reluctantly talked on *Seismosaurus* to interested scientists of Los Alamos National Laboratory in New Mexico. Several of the Los Alamos scientists, geochemist Roland Hagan among them, eagerly forwarded suggestions for finding more of *Seismosaurus* using technologies never before applied to fossil prospecting.

Dr. Alan Witten of Martin Marietta Energy Systems (which operates the Oak Ridge laboratory for the Department of Energy) had devised a computer-based method for creating high-resolution sonic images of buried objects. The system has been used successfully to look for old, buried hazardous-waste containers, and various federal departments want to use it for finding enemy tunnels, coal seams, and buried ammunition. Witten's device has found objects as small as one-third of a cubic yard over a distance of more than one hundred feet.

Hagan read of the device and had dinosaur bone samples sent to Oak Ridge. There, researchers found that sound moved through bone fossils at four thousand meters a second, ten times faster than it did through sandstone, suggesting sound waves could identify dinosaur bones below ground.

Hagan fired shotgun shells into the ground at the *Seismosaurus* site and picked up their subsurface sound on a hydrophone, which registered differences in sound speed as the waves passed through sandstone or dinosaur bone. The results were not conclusive.

A second technique tried at the dig was ground-penetrating radar, used by geologists (including the appropriately named George Jurassic, a University of California San Diego geophysicist) to find faults hundreds of feet below the earth's surface. Los Alamos and Sandia Laboratories crews tried out a lawn-mower-sized sensing unit that emitted pulses every five one-billionths of a second as it moved across the site. The radar penetrated to a depth of ten to twenty feet, sending out a broad, unfocused radio beam. As with Witten's device, bone fossils reflected the sound in a different manner than sandstone, but the complex strata made the radar return signals hard to interpret.

Uranium concentrated in Morrison dinosaur fossils may also help in locating them below ground. Los Alamos scientists used a liquid-cooled gamma-ray spectrometer (usually employed by the lab to detect radiation leakage) to search for the uranium in fossils. But some thirty feet of sandstone overlay the *Seismosaurus,* and Hagan didn't think the spectrometer could find traces of uranium at depths greater than two feet.

The bones also contain, to the researchers' surprise, small amounts of magnetite. This trace mineral was detectable by a sensitive magnetometer that the lab had constructed for testing the magnetic field around a large spectrometer. Bones slightly affect the resistivity in the earth's magnetic field. Magnetometers have been used with some success by University of California/Berkeley archaeologist Kent Weeks in locating below-ground structures around ancient Egyptian tombs in the Valley of Kings in Thebes.

Dinosaur bone also fluoresces under ultraviolet light. A nighttime ultraviolet search of the hillside below the *Seismosaurus* tail vertebrae failed to turn up any dinosaur remains, a positive sign to Gillette that the animal's delicate bones had not been scattered and that more of it might remain within the mesa.

None of these ambitious methods has yet proved fruitful, though Gillette was at once openly hopeful about the prospects and close-mouthed about the results. But they all may yet have applications for digging dinosaurs.

What has proved a highly worthwhile collaboration between Gillette and Los Alamos scientists are chemical tests of dinosaur-bone properties to determine the processes of fossilization. Heidi Schwartz, a geologist and paleontologist at the University of California at Santa Cruz, and a guest scientist at Los Alamos, says some of her group's analytical findings are "overturning popular beliefs of the public and some paleontologists that all bones turn to stone." She and others found when they looked at the microscopic structure of *Seismosaurus* bone, "the pores fill in, but it's not true that bone is petrified. Ninety-nine percent of bone remains bone in fossilization." Schwartz and her colleagues are looking into how particular environments affect what fills in the spaces.

With more conventional fossil-hunting methods, Gillette and his crews have dug and blasted tens of tons of hill away, to expose the entire hind end of a dinosaur, from the middle of its tail to the middle of its back. Two neck bones and some bits of tail that washed downstream millennia ago have also been recovered.

Unlike Jensen's Dry Mesa site, this one was not a jumble of bones from many species, but rather the largely intact remains of one animal. How could something so large and complex be so well-preserved? Geologist Kim Manley, a guest scientist at Los Alamos National Laboratories who has worked on the geology of the site, examined the sediments around the *Seismosaurus* and concluded, "The *Seismosaurus* is buried in stream sand. It slumped down into the water and got buried rapidly, in the next few years, so that it avoided being picked apart." To Manley, the absence of fine-grained clay in the matrix around the dinosaur indicated that a stream, of sufficient volume and force to sort its contents and carry clay far downstream, had flowed over *Seismosaurus* shortly after its death.

As the full extent of the intact skeleton was not evident until several summers of digging, Gillette at first doubted he'd found a dinosaur of historic proportions. It was not until the summer of 1986 that Gillette had the time to study the *Seismosaurus* vertebrae in detail. As he did detailed measurements, and compared the vertebrae with those of *Supersaurus* lent him by Brigham Young University, Gillette realized "just how big an animal this was."

Just how big is debatable. Gillette says, "The standard answer is, in excess of one hundred and ten feet. I say it is the longest dinosaur known, that it is one hundred and twenty-five percent the length of the longest known dinosaurs according to the tail vertebrae. We don't have enough information to say absolutely the length; it could be from seventy-five to eighty-five feet up to one hundred seventy feet. I think one hundred forty feet to one hundred fifty is a fair estimate. We do know comparable elements to other diplodocids [*Diplodocus* and *Supersaurus*], and those of *Seismosaurus* are significantly larger."

Only after he'd established to his satisfaction that his dinosaur was the longest yet measured did Gillette coin the name *Seismosaurus.* One year and three months after he first excavated, he was confident enough to go public with the find. Though scientific protocol calls for announcing a discovery via a formal description, another five years passed before Gillette published one (in late 1991), naming *Seismosaurus.* The enormous cost, labor, and public interest in *Seismosaurus* all led Gillette to go public with his record-setting find long before he was able to analyze it fully.

Gillette has found that *Seismosaurus* is different, in other ways besides its unequaled size, from other sauropods. *Diplodocus,* with a similarly long tail supported by chevron-bottomed vertebrae, would appear to be its closest relative. But *Seismosaurus* has paddle-shaped chevrons, while *Diplodocus*'s are straight in profile. In its overall gestalt, *Seismosaurus* is also singular. Its pubic bone is stubby, suggesting stubby legs, "a very long, very heavy, but very short-legged animal," says Gillette. He opines that the stubby build was an adaptation to support *Seismosaurus*'s singularly enormous bulk.

With the publication of a preliminary description, Gillette's work on *Seismosaurus* is far from done. Preparation of the *Seismosaurus* has not proceeded rapidly. Fossils so huge, fragile, and difficult to discriminate from their rock matrix require thousands of hours of preparation. Those hours require funding resources Gillette does not possess. He and BYU researchers were turned down for a NSF grant to study their giant animals, in part because none of them was a veteran dinosaur scientist. Gillette has raised only a small fraction of the $1 million he estimates it would take to fully excavate and prepare the *Seismosaurus*.

For now, the *Seismosaurus* belongs, in title, to the New Mexico Museum of Natural History. Portions of the dinosaur are being worked on in an Albuquerque warehouse. Gillette has more in his offices, and BYU still more at its lab. Cleaning the bones already found will take at least five man-years, at $25,000 a year, money Gillette doesn't have.

But even with funding, Gillette anticipates that *"Seismosaurus* is a project that will last me the rest of my life. I just hope to God I never find another giant sauropod."

Is *Seismosaurus* truly the longest dinosaur? *Ultrasaurus* the biggest? Jack McIntosh of Wesleyan University, the only paleontologist to attempt to thoroughly organize sauropods, is the person best qualified to judge the assertions of Gillette, Jensen, and any other claimants to the distinction of having found the world's largest dinosaur.

McIntosh has examined all the bones in question. He was one of the first outsiders at Dry Mesa after the giant pelvis was found. He has worked with Gillette and has long been convinced of the worthiness of Gillette's claim that *Seismosaurus* is a new and singularly long animal. Of *Ultrasaurus,* he says, "It's known from so few bones. I'm skeptical." McIntosh is not convinced *Ultrasaurus* is something new. Like other researchers, he thinks it possible that *Ultrasaurus* might be just a big *Brachiosaurus.* Its huge vertebra with a high spike, used to name the animal, might belong to the rear end of a diplodocid. Nor did the shoulder bones excavated at Dry Mesa provide clear evidence for *Ultrasaurus* or *Supersaurus* in McIntosh's mind. "But I thought all three were different." Still, it's possible, says McIntosh, that *"Supersaurus* and *Ultrasaurus* may be the same animal." As for *Supersaurus,* McIntosh is more confident of its identity, but not utterly convinced. "I'd say it's a diplodocid with a two-thirds' chance of being new. But I hate to speculate. This field has gotten into more problems just that way."

If Jensen's papers failed to establish unequivocally the singular nature of his finds, including the biggest-dinosaur claim, neither did the fossils themselves. But they may yet. When he retired in 1983, Jensen left behind a staggering one hundred tons of jacketed fossils, white-plastered boulders full of dinosaur bones so numerous and voluminous they fill not only an Agricultural School warehouse but the basement of the vast BYU football stadium. They lie unopened to this day, and no one, Jensen included, seems sure what is in them. BYU remains disastrously short on funds for paleontological research. It takes two hundred hours of preparation to clean a single one of those bones, and in 1991, the university laid off its last fossil preparator.

So, Jim Madsen, Gillette's predecessor as Utah state paleontologist, considered Jensen's excavating extravaganza a colossal mistake. "He'd boast of collecting one hundred tons of dinosaur bones in matrix. That's a travesty. You can't do anything with those bones now. There was a lot of work to do at the quarry. He wanted a major collection of dinosaur bones. Publicity is fine, but you need to get them into a prepared collection, that's the real challenge." (Oklahoma Museum of Natural History curators have done just that with their own dinosaur fossil collections, including remains of what McIntosh thinks might be the largest *Apatosaurus,* working for several years from a $150,000 National Science Foundation grant.)

Jensen's mountain of plaster-covered finds may include a dinosaur bigger than any he or Gillette has named, or any yet known from fossil evidence. In 1985, Jensen found one-third of an extremely large sauropod femur in a uranium miner's front yard in southern Utah. The head of the femur was a whopping five feet six inches in circumference, making it, comparatively, the largest bone Jensen had ever seen. He did not name it. Nor is he inclined to help others sort out the still vexing problems of *Supersaurus, Ultrasaurus,* and his other finds as yet unnamed. "They don't do things the way I would. I just had to leave it all behind and not look back."

For David Gillette and Jensen's successors at Brigham Young University, preparing and understanding the three largest dinosaurs will be a career task. But for other paleontologists, the evolution, behavior, and the environment of those and other huge sauropods are of greater scientific concern. And from *Seismosaurus,* scientists are getting some answers to their questions.

Jurassic World

Vast sauropods and plate-backed stegosaurs lumber across a swampy land, stalked by the terrifying *Allosaurus* (see page 144). This is our classic image of The Land Before Time, a dinosaur world snatched from the cusp of two dinosaur times, the Late Jurassic and Early Cretaceous—some 145 million years ago. Those dinosaurs were unquestionably real—we've known them for more than a century from the American West. But their miasmic environment is fiction.

The Jurassic period, 208 million to 145 million years ago, was a time when the earth's surface broke apart dramatically. Seams that had torn at the Triassic-Jurassic boundary in the North Atlantic in New England and Nova Scotia 200 million years ago split open elsewhere along the ocean's present boundaries, creating deep rift valleys. The North Atlantic began filling in one such rift 165 million years ago, and the South Atlantic did so 40 million years later. The great shallow sea to the east of Pangaea, the Tethys, spread westward, dividing the supercontinent in two. Soon thereafter, by geologic standards, Laurasia split in two and Gondwana split away from Europe and into many fragments. The Turgai Sea rose in the Jurassic to make another continental separation, dividing Europe from Asia.

The Jurassic climate was warm and for most of the year dry, favoring the conifers and cycads (resembling palm trees and closely related to ferns) that had begun to dominate in the first dinosaur era, the Triassic. The major forms of Jurassic trees were probably not fast-growing nor particularly rich in foliage, yet they supported the largest animals ever to walk the earth.

Among the ascendant dinosaur forms in the middle of their era were creatures great and small—large carnivores such as *Gasosaurus,* which lived at Dashanpu, and the hypsilophodonts, such as Dashanpu's

Xiaosaurus, small to medium-sized (six to sixteen feet long) plant-eaters. By the end of Jurassic time, the hypsilophodonts shared the roughage with larger browsers (sixteen to thirty-three feet long), the iguanodonts. And both must have made way for chomping giants— relatively slender, whip-tailed forms such as *Diplodocus* and *Seismosaurus* among them, and stouter creatures such as *Brachiosaurus, Ultrasaurus,* and their kin.

These were almost unimaginably huge animals. They attained weights of many tens of tons and lengths in excess of a hundred feet. Even the smallest sauropods were more than twenty-five feet long, larger than nearly all other dinosaurs. By late in the Jurassic, nearly identical forms of all these browsers had spread from Africa to Western North America. But they are known best from the Morrison Formation of the western U.S., a 1-million-square-kilometer area of sedimentary rock now exposed in Utah, New Mexico, Texas, Oklahoma, Colorado, Wyoming, South Dakota, and a bit of Montana as well. These sediments were laid down late in the Jurassic and early in the Cretaceous era, from 150 million to 130 million years ago.

Some of the dinosaurs of this ancient American West were tiny. But this was also home to titans. Allosaurs twenty feet in length have been found in abundance in central Utah. *Stegosaurus,* thirty feet long and the largest of the plate-backed dinosaurs, has been known for a century from Wyoming and other states. And in 1991, paleontologists in Colorado unearthed the earliest North American example of nodosaurs, armor-sided tanks a ton in weight and twelve feet long. But the dominant dinosaurs in the Morrison Formation are those truly monumental dinosaurs, the sauropods. As throughout the world, western North America was titans' turf.

The American West had been underwater in the Middle Jurassic. But by 145 million years ago, the Jurassic seas had retreated northward into Canada for the fourth and final time in that epoch. The record that remains of the Late Jurassic and Early Cretaceous environment of the American West is not, however, one of swamps or of deep lakes, as scientists long believed, but of a harsher land. The fossil record is curiously rich in dinosaurs, poor in other life-forms. Few small aquatic vertebrate fossils and little coal (from decayed plants) remain, suggesting that what large bodies of water existed were too hot or salty to support much plant or animal life. Perhaps they were large playa lakes, seasonal sheets of alkaline water covering the flat portions of basins.

The most habitable terrain for dinosaurs was a mosaic of lakes, rivers, and a vast floodplain below the highlands of what is now the Great Basin of Nevada, Arizona, and western Utah. With watering holes and leafy oases few and far between, the giant dinosaurs likely migrated throughout the American West. That's what Peter Dodson, Kay Behrensmeyer of the Smithsonian, Jack McIntosh, and Bob Bakker concluded in their pioneering 1980 study of Morrison environments.

Without swampy land, there was no place for water-loving, sluggish sauropods. Instead, the giants would have to have been landlubbers, and highly mobile ones at that. The evidence for wandering dinosaur herds included the observation that the Morrison Formation is "unique in the existence of multiple concentrations of bones" of the same dinosaurs over sites many miles apart. Footprints recently discovered in South Korea appear to show calf-sized young sauropods milling about, suggesting young sauropods were tended by a herd of adults.

Dodson compares the Morrison scene of 145 million years ago to the Serengeti plain in modern Africa. Grasses hadn't yet evolved, but there were cycads, conifers. There must have been trees by watercourses in abundance to feed the giant dinosaurs, even if environmental conditions haven't favored preservation of them. "The dinosaurs wandered like wildebeest," Dodson surmises, "following the rains and the cycles of plant productivity."

Today this region is a dry and largely barren plain, interrupted by a spectacular badland terrain of towers and canyons. What remains of the life in mid-dinosaur time are tracks and traces, but awesome traces they are.

From the remains of *Seismosaurus* and its surroundings, geologist Kim Manley compiled a portrait of the world in which *Seismosaurus* lived and died. "The were large alkaline playa lakes with no apparent drainage." A number of volcanic eruptions to the west made them even more uninhabitable, showering them with ash. Uranium, soluble in water, percolated up and into bones washed into the lake bottoms, making them radioactive. (Indeed, one way to spot Morrison Formation dinosaur bones in the West or in museums is with a Geiger counter. Hear them click—the bones are "hot.")

Harsh as these conditions were, Manley notes, the environment could be "very pleasant in places, like parts of the Serengeti," a conclusion very much like that drawn for all of the Morrison Formation by Bakker, Dodson, McIntosh, and Behrensmeyer a decade before.

Manley's detective work also extended to previously unknown and still unseen details of *Seismosaurus*'s anatomy. Manley found hundreds of smooth, highly polished stones up to four inches in diameter around the site. Immediately she noted, "They're the wrong rock types for the area, they must have been brought in."

What brought them to that site? Sauropods, in their bellies. Gastroliths are stomach stones, used by animals, such as the grit-swallowing chicken, to grind plant food. These rocks had been found in the abdominal cavities of fossils of plesiosaurs and suspected for dinosaurs, but never definitively noted within the body cavity of a fossil dinosaur.

Within the articulated rib cage of *Seismosaurus,* Gillette has discovered 230 gastroliths, "a bushel-basketful," as he describes it. The stones, most highly polished, range in size from a walnut to one as big as a grapefruit, with most smaller than a peach. The big stone, swallowed inadvertently, may have choked and killed the *Seismosaurus,* Gillette speculates. The stones were discovered in two clusters, one fore and one aft in relation to the beast's stomach. The alignment of stones suggested one batch belonged to a predigestive crop and the latter to an intestinal gizzard.

Sauropods had famously small brains, within relatively tiny heads. Their spoon- or pencil-shaped teeth look hardly sufficient to thoroughly masticate huge quantities of tough vegetation. Digestion within a stomach vat would be greatly aided by strong muscular contractions, grinding a bellyful of millstones against a half-chewed mass of leaves.

Gillette was dubious at first of the identity of the stomach stones. But their location within the animal's skeletal frame, and the clearly exotic nature of these crystalline volcanic riverstones, convinced him he had gastroliths.

Understanding Sauropods

With whatever anatomical and behavioral adaptations they developed, from migrating patterns to gizzard stones and heat-producing fermenting vats (see Chapter 6), sauropods triumphed mightily. Sauropods are known in one form or other from the Late Jurassic of Europe, Asia, and North and South America. In parts of the world, particularly the southern continents, sauropods prospered to the very end of dinosaur times. And the biggest, writes paleontologist Peter Dodson, "clearly approached the theoretical maximum for mobile terrestrial organisms."

Seismosaurus may have uniquely explored the anatomical limits of life on land, but it is still only the largest of the large, one more remarkable success within a group whose longevity, diversity, and distribution are unexcelled among dinosaurs. Sauropods are known from the Early Jurassic to the Late Cretaceous, a span of more than 130 million years. There are more kinds of sauropods (ninety genera named) than of any other group of dinosaurs. And they are found from every continent.

Yet they remain among the least understood of all dinosaurs. Their behavior and evolution, and much of their anatomy, is a mystery, their taxonomy a mess. Good sauropod skeletons are hard to come by. Of the fragile and highly diagnostic sauropod skulls, we have fewer than twenty worldwide, Dashanpu's treasures notwithstanding. Sauropod eggs are also rare, as are the bones of juveniles. The reasons for this scarcity are unknown. Perhaps, as Dodson and others conjecture, juvenile sauropods had delicate, unfused bones that were therefore poorly preserved. And among adult sauropods, only five kinds are based on reasonably complete skeletons, a phenomenon that appears related to the animals' enormousness. Says Dodson, "Once they get over thirty feet, you don't get the whole dinosaur."

Nonetheless, paleontologists have succeeded in identifying what sauropods have in common, if not which are genuine and how all are related. *Sauropod* means "lizard feet," but aside from having five toes, sauropod feet weren't very lizardlike. They could be better compared to an elephant's. Both have broad stubby toes, apparently backed by a thick wedge of tissue that supports their body with each footfall. Sauropod feet were distinguished, however, by large claws on the first three toes of each hind foot, and by a large claw on each of their "thumbs."

Sauropods were in many ways an archaic form of dinosaur, with simple spatulate or pencil-like teeth, and the smallest brain-to-body ratio of any dinosaurs. They look oddly pinheaded in proportion: behemoths with heads the size of those of modern horses. Working with brains smaller than those of house cats, they were probably incapable of much sophisticated, learned behavior. Yet they were enormously widespread and long-lived successes.

Lately, cladist interpreters have identified just what makes a sauropod, defining many of their shared characteristics. Jacques Gauthier names thirty such common features, among them a small skull and ten or more neck vertebrae.

Since so few skulls are known from all sauropods, backbones offer the best present hope of telling one sauropod from another, as Jose Bonaparte has lately tried to do. Perhaps because they functioned as weight-saving and weight-bearing structures, these bones have taken on many architectural peculiarities in sauropods, including variously shaped holes and knobs. Presumably these features evolved to increase the strength while reducing the weight of the vertebrae as the animals achieved ever-increasing size.

Telling one sauropod from another by their backbone, pelvis, or shoulder blade architectures, or any other clue commonly used to I.D. dinosaurs from teeth to skull shape, is a difficult business indeed. "You have to be a masochist to take them on," says Johns Hopkins University paleontologist David Weishampel. Aside from the sketchiness of their remains and the muddle of their systematics, there's one more fundamental problem with studying sauropods. Weishampel notes, "They are just so damn big, you either have to be an acrobat or very strong."

Jack McIntosh has been an acrobat and strongman at times, both unlikely roles for a small, portly gentleman of advancing years and receding hair. An owlish, bespectacled man in his sixties, McIntosh speaks reluctantly, and then loudly, at a rapid clip. He dislikes publicity and disdains even collegial praise. But McIntosh's bark is far worse than his bite. He has kind words for most everyone who has ever worked with sauropods, though he knows well that most have, in one way or another, gotten them wrong.

If McIntosh himself has made mistakes, it isn't for want of effort. From China to Argentina, and in every museum collection of significance in between, he's examined sauropods firsthand. "He's the most underrated dinosaur paleontologist," says Bob Bakker. "He's helped everyone and he won't take any credit." Jensen acknowledged McIntosh's contribution to his work by giving his *Ultrasaurus* the species name *macintoshi*. "My first memory of him," recalls Bakker fondly, "is walking into the basement of the Peabody Museum and seeing this little man way up in the air, tugging at some enormous bone." Museum basements are a good place to find McIntosh, for, as Weishampel points out, "if you want to look at sauropods you've got to go to places like hot, dark rooms in the back of the American Museum and climb narrow scaffolding and ancient pallets."

More often, McIntosh works in a modern tower on the edge of the ivied Wesleyan University campus in Middletown, Connecticut. Volumes of *Physical Review* dominate the shelves of his long, narrow office; McIntosh is a distinguished physicist with a professional interest in nuclear theory.

Tucked away in one corner, however, are indications of McIntosh's avocation as a sauropod anatomist: dinosaur volumes and several sauropod teeth. Sauropods came first for McIntosh, a fascination that began in his suburban-Pittsburgh childhood when he often viewed the fossils at the Carnegie Museum. Though he'll happily expound for hours on details of sauropod anatomy, he condenses his personal dinosaur history to a few clipped words. "I just happened to be always interested in dinosaurs. I had already studied sauropods before I came to Yale [for undergraduate and graduate studies]. When I got there, I worked in the Peabody Museum for scholarship money. The day I arrived, they were putting together *Apatosaurus* vertebrae. The museum happened to have an excellent collection of sauropods, and I liked their bigness, so I studied them, just for something to do."

The obstacles to understanding sauropod anatomy and evolution never impeded McIntosh. "Eventually we will fill them in." Nor did the experience of decades of working with sauropod bones discourage him, as it has other paleontologists. Yet he acknowledges the special problems studying sauropods presents. "Sure they're heavy as hell," he barks, "and you can't send them through the mail. With sauropods you need two men to move one big vertebra. A limb bone can weigh hundreds of pounds. If a femur falls over, you'd get killed." McIntosh has avoided serious injury to date, but allows, "I stove all my fingertips in at one time or another."

If McIntosh's name is known at all outside his scientific fraternity, it is for solving the case of mistaken identity of the familiar sauropod "Brontosaurus," whose genus name, as McIntosh has pointed out, is properly *Apatosaurus.* In 1975 he and colleague David Berman of the Carnegie Museum straightened out some literally wrongheaded scientific thinking that had stood for a century. Many museums, the American Museum of Natural History among them, had long followed the lead of nineteenth-century dinosaur paleontologist O. C. Marsh of Yale, who'd stuck a short and blunt-snouted head on his creature, a head that had been found four miles from the "Brontosaurus" body. From examination of the fossils and early records, McIntosh and Berman realized Marsh had given the animal the head of another sauropod, *Camarasaurus.* Following McIntosh and Berman's investigation, "Brontosaurus" was given its proper head, a narrow, big-eyed, and pencil-tooth diplodocid-style skull.

McIntosh and Berman's reexamination of "Brontosaurus" was part of a much larger effort to regroup sauropods by their anatomical similarities. McIntosh has established five, possibly six, families of sauropods, with several others still too little-known to identify further.

The most massive of all were the brachiosaurids (more than fifty-five tons and one hundred feet long), to which *Ultrasaurus* probably belongs. Like a giraffe, a brachiosaur towered over its contemporaries, with its shoulders, surmounted by a long neck, elevated well above its hips. Its nostrils were placed high above its eyes and low, broad snout.

This suggested to early paleontologists it lived largely submerged. To modern scientific speculators the nostril placement suggests it may have had a trunk.

Supersaurus and *Seismosaurus* are likely members of the diplodocids. Dinosaurs of this family have slender nostrils set above eyes far back in the skull, peglike buck teeth better suited to stripping succulent plant fronds or even straining aquatic "gruel," as Dodson speculates.

Diplodocids may have reached high into trees, using hind legs and tail as a supporting tripod. Oregon artist Mark Hallett, a consummate dinosaur painter, has theorized that diplodocids had a powerful bite. He contends they ate mostly conifer needles, leaves, and cones, and that their nostrils were set far back in their heads as a result of an elongation of the skull to accommodate strengthened jaw muscles. Hallett also speculates that diplodocids fed on underwater plants. Nostrils atop their heads allowed them to breathe while drinking and feeding underwater.

Tooth wear studies by Carnegie Museum paleontologist Tony Fiorillo support Hallett's view of diplodocids as tree-top feeders. Fiorillo found a finer pattern of wear on the pencil teeth of *Diplodocus* than on the spoon teeth of *Camarasaurus.* Fiorillo suggests the difference in wear may reflect a grittier diet of lower-story plants consumed by *Camarasaurus* while *Diplodocus* dined on more delicate tree-top foliage. Whatever their diet, the Late Jurassic sauropods grew longer than any animals before or since. Some, such as *Diplodocus,* were slim-bodied with whip tails and snake necks, and so weighed less than other sauropods half as long. Others, such *Apatosaurus,* were more heavily constructed.

Few paleontologists are inclined to challenge McIntosh's organization of sauropods. None have the stamina or the resources to investigate these animals as fully. Others will continue finding new giants. In the summer of 1991, Museum of Western Colorado paleontologist Harley Armstrong announced that he had found the backbone of the largest apatosaur yet known. But precisely where the giant sauropods, particularly *Seismosaurus, Ultrasaurus,* and *Supersaurus,* fit into the new ordering of sauropods is a matter not even their discoverers agree upon. And those who work on sauropods can still dream of finding, or having found, the largest dinosaur of them all.

Selling
Fossils

Dinosaur fossils are a seller's market, and a lucrative one. Bids from Japanese and German collectors for the newly revealed and privately held *Archaeopteryx* specimen in Solnhofen are rumored to approach $5 million. Says Gillette, "I know of one stegosaur in private hands that a guy is selling for $250,000 unprepared, and twice that to prepare it." Dinosaur fossils are hawked each February at the Gem & Mineral Show in Quartzite, Arizona, near Tucson.

What's wrong with a little dinosaur trade? Fossils were traded as objects of beauty before they were ever studied. Physician Gideon Mantell, who described the first fossils, later identified as a dinosaur's, sold his collection to the British Museum. Selling fossils remained, for a century, a common practice among paleontologists. Now, however, the stakes are higher. And paleontologists want themselves and their potential objects of study out of the market.

For one, notes Gillette, "commercial trade does a tremendous disservice to the academic world, jacking up the prices out of the range of many institutions and taking the fossils out of the public domain." Many American museums will not buy fossils from fossil collectors. Instead, says Gillette, "the fossils end up in some broad gray zone, sold to foreign museums and individuals with money."

What's more, commercial collectors often remove fossils without the care for the bones and the study of their environmental context that professionals routinely provide. And they may raid existing scientific research sites, making off with or even destroying valuable study materials. In the 1950s when uranium was in high demand, uranium hunters would grind whatever dinosaur bone they came upon. Jensen lost part of a small herbivore he identified as *Hypsilophodon* to rock or fossil hunters. He'd brought the specimen back to BYU for study, realized upon preparation he might have left part of the animal in matrix at the Colorado dig site, and returned there only to find the quarry scavenged and bones removed. As he told a reporter, "We may never find another *Hypsilophodon,* and we'll never know what it looked like."

In the summer of 1991, Swiss commercial collectors were extracting a rare juvenile *Allosaurus* from a Wyoming site, when Bureau of Land Management Officials discovered that the specimen was on federal, not private, land. The BLM confiscated the fossils and assigned the Museum of the Rockies to excavate and prepare the remains.

Rights to many of Museum of the Rockies paleontologist Jack Horner's most promising sites have been purchased from landowners by commercial collectors. Other sites of Horner's have been looted by fossil "rustlers," who made off with dinosaur bones of scientific import.

Much commercial dinosaur-collecting is quite legal. Dinosaurs found on private land are fair game for sale, and since their origin is difficult to trace, many illegally obtained from public lands are pawned off as privately obtained. And while many cultural artifacts, down to the commonest arrowhead, cannot legally be surface-collected from federal and state lands, fossils can be, providing the collector has obtained a permit. (Utah has passed a state law making such collecting illegal on its lands).

The National Academy of Sciences sought to address the fossil-collecting problem in a 1987 set of recommendations. These were drawn up by a thirteen-person panel, consisting of vertebrate and invertebrate paleontologists, a paleobotanist, lawyers, and businessmen. Their conclusion, that "commercial collecting of fossils from public lands should be regulated to minimize the risk of losing fossils and data of importance to paleontology," did not satisfy vertebrate paleontologists. For one, no distinction was drawn between invertebrate fossils, which are much more common, and vertebrate fossils. Oftentimes, with dinosaurs or extinct fish, the specimen collected by a commercial collector may be the only known example of a species, invaluable to science yet perhaps unavailable to it.

SVP paleontologists have asked for a specific federal prohibition of commercial collecting on federal lands. Gillette has been assiduous in attempting to apply Utah laws against fossil-collecting. Yet, ironically, dinosaur bones are sold, in Utah, by the professional paleontologists who collected and prepared them, at Brigham Young University. There, in the tiny museum gift shop adjoining the preparation labs, tiny scraps of dinosaurs, bits chipped away or left over in the cleaning and reconstruction of skeletons, are sold for a quarter. Not much of an operation, "but it helps bring a little money into the museum, and they're bits we'd have thrown away otherwise," maintains BYU curator of paleontology Wade Miller. Miller sees the entire fossil-sale issue as complex. "It's a sticky business. Many commercial collectors contribute a lot to science." Still, BYU's own violation of SVP guidelines has some paleontologists, Madsen and Jensen included, upset. "I never sold fossils" says Jensen. Adds Madsen, "I don't think BYU should be selling it. If I had a museum, I wouldn't do it. People see professionals trading in dinosaur bone, they think it's all right for them."

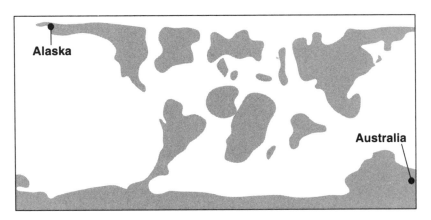

The breakup of the supercontinents, 113–65m. yrs. ago

T   J   C

**113-65m.
yrs. ago**

245m.   65m.

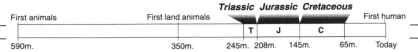

AGE OF THE DINOSAURS

Triassic Jurassic Cretaceous

First animals   First land animals   First human

T   J   C

590m.   350m.   245m. 208m.   145m.   65m.   Today

Dinosaurs at the Poles

A dozen animals, none larger than a wallaby, tilt forward on their long hind legs to nibble on greenery along the fringe of a cold lake. The pastoral scene is not an uncommon one for southern Australia in midwinter.

The time, however, was 110 million winters ago. These were dinosaurs, not marsupials. And their habitat looked nothing like any in Australia today—a great rift valley from which thrust huge granite tors. Weedy vegetation dominated in the valley, interrupted by braided streams, this and other lakes, and patches of ginkgo forest. The weather was decidedly cool, dropping below freezing long enough and often enough to leave a crust of ice on the surface of the lake.

Though grazing largely in open terrain, the dinosaurs would have been hard to glimpse, even for each other. Save for moonlight, the valley was dark. It had been dark for six weeks and would be dark for six weeks more. The little dinosaurs' owlishly large eyes were opened wide to catch a shadowy glimpse of each other, or the approach of a hungry allosaur eager to make them into lunch.

This Australia was not an island, but one corner of a vast polar land mass. Its creatures lived much farther down under, well within the antarctic circle (between seventy and eighty-five degrees latitude).

Considerable imagination is required to picture such a place in time, for there is no equivalent environment today—cool, moist, and temperate, a land of evergreen plants and trees and few flowers, its rivers flooding in spring, lakes freezing in the weeks or months of continuous darkness each winter. Today, on land this close to the South Pole, one can find only lichens, seals, and penguins. Back then, it was as if an Oregon forest prospered in northern Alaska.

The dinosaurs discovered in this ancient Australia represent not only a new and strange assortment of animals, but along with those lately found in arctic Alaska, they constitute compelling evidence that many dinosaurs coped with seasonal cold and dark. These conditions were harsh, not unlike the Nuclear Winter forecast by some scientists as the likely aftermath of an exchange of atomic weapons. A similar environmental scenario is envisaged by some catastrophe theorists in the dinosaur-extinguishing aftermath of an asteroidal collision with the earth, 65 million years ago. But of greater interest and puzzlement to paleontologists is how these dinosaurs dealt with winters of cold and dark—by migration, hibernation, warm-bloodedness, or some other adaptation.

Dinosaurs are notoriously hard to come by in Australia. Among continents, only Antarctica has yielded a slimmer fossil record. A century of dinosaur prospecting in Australia had produced some intriguing and beautiful fragments, including dinosaur bone fossilized as opals. But only two dinosaur skulls and three skeletons had been found on the continent (the best specimen being a slender little ankylosaur, an armored dinosaur less than nine feet long, found in 1990 in north-central Queensland).

Yet, over the past few summers, researchers have discovered evidence of a varied dinosaur fauna peculiar to Australia, and one of the fullest records yet of dinosaurs from 135 million to 100 million years ago, the Early Cretaceous era in which dinosaurs are little known worldwide.

The newfound Australian dinosaur fossils did not give themselves up without a fight. The spot in the ancient valley where the little dinosaurs browsed in the Cretaceous, or rather where the nearby stream deposited their corpses on its banks, is now a high, steep cliff-face of green sandstone and mudstone at the southeastern tip of Australia. A wild and handsome place, it is now called Dinosaur Cove.

Early Cretaceous rock exists farther inland from this windswept square mile of scraggly coast, but it is seldom exposed. When it is, low, bushy overgrowth obscures the rock. And even if one were to dig down to it, no bones would appear. Chemical erosion has long ago dissolved the fossils.

Along this stretch of beautiful coast, pounding waves cut away at the rock walls, occasionally exposing fossils. A government geologist, William Hamilton Ferguson, came upon the region's dinosaurs at the turn of the century. Ferguson was mapping the coal-bearing formation eighty miles to the east along the coast when he found what was later described as a toe claw from a carnivorous dinosaur.

For seventy years after the "Cape Paterson Claw" discovery, not another dinosaur fossil was found in all of Victoria State. Then, in 1978, Tim Flannery, a Monash University of Melbourne geology graduate student, and his cousin, John Long, took Ferguson's map and located the site of a dinosaur find at Eagles Nest. Within minutes of searching along the cliff, Flannery had found what proved to be the top of a dinosaur humerus (upper front-limb bone). Flannery spent the rest of the summer looking along the coastal outcrops of the area and found thirty more bones, including an ankle bone with at least two distinctive features of an *Allosaurus*.

The following two summers, a group of paleontologists came looking for more fossils. These included Flannery, Ralph Molnar, and his fellow American émigrés, Pat Rich-Vickers, a Monash University professor, and her husband, Tom Rich, a curator of the Victoria Museum in Melbourne. Given the geological similarity of this formation to that of Dinosaur Cove, several miles to the west, the explorers decided to

Tom Rich takes a break from dynamiting for little dinosaurs at Dinosaur Cove, Australia. [Photo courtesy Earthwatch]

expand their search to the Cove in 1979 and 1980. Progress was slow, not only because of the rarity of the fossils, but also because of the difficult nature of the terrain.

Tom Rich is an adept climber, thin and youthful for a man in his early fifties. Bespectacled and slightly disheveled, Rich is as cheerful and gregarious as the stereotypical Australian. His speech is colored by Aussie slang, spoken with a decided American accent. He'd come on this dinosaur hunt out of only casual interest. Despite an early fondness for dinosaurs (on his twelfth Christmas, Rich read Roy Chapman Andrews's *All About Dinosaurs*), he'd looked in other directions in college, studying computer programming, and physics with Luis Alvarez, before settling on a career in paleontology. He married another Berkeley undergraduate, Pat Vickers, with similar professional inclinations. Both went to Columbia for graduate studies, where Pat's interests were directed toward fossil birds. Tom's were mammals, hedgehogs among them. After Columbia, they resolved to go together wherever either one got the first job offer. Tom was offered a position as curator of paleontology at the Museum of Victoria in Australia while Pat was there on a Fulbright Fellowship. Pat eventually found a post at Monash University in Melbourne, and today she is the better known and more widely published of the two—as a researcher and author on fossil birds and other vertebrates in Australia, New Zealand, and China.

Neither of the Riches had ever worked on dinosaurs when they joined the exploratory team southwest of Melbourne. Investigating the coast, they and other scientists found many fossils, dinosaurs included. Four localities were relatively rich in fossils, including Dinosaur Cove, the richest of the four.

Dinosaur Cove is a lovely spot, tucked away in Otway National Park, on the flanks of the low Otway Range, which runs parallel to the coastline for more than fifty miles. Much of its hills are covered in majestic stands of eucalyptus, with groves of tree ferns in the wetter reaches.

Excavations were undertaken in the exposed rock face of the cove, and a dozen fossils were unearthed at or near the surface in the next two years. But Tom Rich and others found they could dig no farther into the ancient stream channels, now stubborn sandstone, with their picks and hammers. "We either had to tunnel in or give up," Rich recalls. "I wasn't terribly anxious to try tunneling. We thought we could find a better site elsewhere."

They never did. And the Dinosaur Cove explorations might have been abandoned were it not for the public interest generated by a touring Chinese dinosaur exhibit at the Victoria Museum in 1982. Over the next two summers the excavation work continued on a shoestring budget.

Rich's camp above the cove was an informal affair, "an accumulation of tents, ancient mildewed furniture, hemp lines strung to dry wet, soiled clothes, and some anachronistic multicolored flags. It looked like a hobo jungle of the 1930s," recalls Earthwatch volunteer Antoinette Beuche, a Michigan lawyer.

Volunteers built a cable to haul stones up from the cove, donated drilling equipment, prepared fossils, supplied food. Though always hard-pressed for funding, Rich most needed laborers. Work at Dinosaur Cove was tedious. "Mainly it's hard-rock mining, bloody hard," says Rich. Digging away at the fossils from below resulted in the discovery of another two thousand bone fragments (most of smaller vertebrates), and a few whole bones and teeth, from three sites within a quarter mile's distance, each at a different height in the rock wall. The last found, Slippery Rock, proved the most productive, but not quickly.

A rock-drilling company donated its equipment and services to help Rich dig a tunnel at Slippery Rock. With scarcely a week left in the 1986 field season, and little money remaining for the dig, Rich's team found the partial skull of a dinosaur. A frantic search soon produced the missing pieces. The next day two pieces of vertebrae, one associated with a partial hind limb, were turned up. The fragment of the skull roof that "had been peeled off like an orange rind" during excavation, as Rich later described it, bore a clear imprint of the top of a brain. Though smaller than a lemon, on an animal as big as a grown woman, this was a huge brain for a dinosaur, with especially large optic lobes.

Rich's reaction to the discovery was not, he recalls, one of elation. "To keep myself going when finding so little for so long, I never let my emotions soar with false hope. It was months before the discovery caused any feelings."

Dinosaur Cove is a wild and scenic stretch of southern Australian coastline.

Rich did immediately note that several features of the skull were, to him, highly unusual. The structure of the teeth, overlapping in a massive jaw and lacking in canines, suggested the animal was a hypsilophodont, one of the small bipedal, herbivorous dinosaurs known from a few other scattered Australian finds, and from other continents as well. Rich speculated initially that the fossils might belong to a new hypsilophodont, or to one of the first of the bigger, also successful herbivores, iguanodonts, known elsewhere in the world from earlier deposits in the beginning of Cretaceous times. While the teeth on the newly discovered skull were primitive, the braincase was not. It was sufficiently roomy to hold a brain three times that expected for a reptilian brain. It was, in fact, a bigger brain in comparison to body size than had been known for any dinosaur, except for *Troödon* and its close-cousin theropods.

Hoping to find more of this or other dinosaurs at Dinosaur Cove, and fearing that the sea might wash away this fossil layer, Rich extended the digging season a week. "Otherwise, the layer would have been lost forever," Rich recalled.

By the end of another lengthened field season in 1987, Rich and his crew had found several more isolated dinosaur bones, partial skeletons of four or five fish, and a pelvis of a dinosaur perhaps five feet long. Small by dinosaur standards, it was nonetheless larger than any other dinosaur found at Dinosaur Cove.

Previous finds from the area had revealed more than one hundred and fifty different inhabitants of this environment, including plant fossils and pollen, and pieces of pterosaurs, plesiosaurs, lungfish (which today reside only in Australia, tropical Africa, and South America) and primitive bony fishes, turtles, small hypsilophodont-type dinosaurs, a small *Allosaurus,* and oddest of all, a large carnivorous amphibian, a labyrinthodont, the youngest of its kind by 50 million years.

In Rich's own quarry, some six hundred specimens were recovered that summer, about half the previous year's haul. These were of far better quality, however, because, found in clay stone, they had come to rest in very quiet water and were little disturbed. Previous finds, from sandstone rich in clay balls, had been washed to the bottom of streams. There, skeletons and individual bones were broken apart by fast-flowing water.

With the discovery of fossil-rich clay stone, Rich left Dinosaur Cove in 1987 optimistic that he might find more dinosaur skeletons at the site. However, analysis of the new and peculiar finds already made took precedence, and Rich spent the summer of 1988 not quarrying, but traveling and researching, comparing Dinosaur Cove fossils with femurs, hips, teeth, and skulls of other small dinosaurs, and discussing the dinosaurs and their Lower Cretaceous environment with others who had studied them in Australia and elsewhere.

In early 1989, the excavations produced 1,300 more fossil specimens, including one, and possibly two, hind limbs of a small dinosaur, only the second instance of articulated dinosaur bones from Dinosaur Cove.

The leg bones are distinctive for yet another reason: they show evidence of a severe fracture, healed over, during the dinosaur's lifetime. This kangaroo-sized hypsilophodont had clearly lived years after an injury from which a modern kangaroo, according to the Riches, would be unlikely to long survive. The find suggests to them these animals lived, and moved, in ways quite unlike the kangaroos they superficially resemble. Perhaps, though the Riches do not suggest it, fellow dinosaurs cared for their injured comrade, as has recently been shown to have occurred among saber-toothed cats.

The Australian Connection

In late 1988 and early 1989, Tom Rich and Pat Rich-Vickers published their conclusions about Dinosaur Cove dinosaurs, identifying them as belonging to three genera of small to medium-sized hypsilophodonts. Two of the dinosaurs were new. The big-brained creature they named *Leaellynasaura,* honoring by the genus name their daughter, Leaellyn, who at the age of two, long before Tom Rich had begun digging dinosaurs, had asked for a dinosaur of her own.

The dinosaur she got was also a child. While *Leaellynasaura's* skull had tooth sockets, it had no teeth, nor evidence that teeth had ever occupied the sockets. That was a significant piece of negative evidence, for it suggested the skull belonged to a juvenile. So, from their size, do all but two of several femurs ascribed to the dinosaurs from Dinosaur Cove. Those leg bones, some smaller than an inch long, came from young animals as small as a twelve inches in length.

Even in adulthood these dinosaurs were small, "not much bigger than a turkey," says Rich, and weighing perhaps less than twenty-five pounds. The single large femur, oddly twisted, and nearly four times the size of the others, may well belong, the Riches believe, to another as yet unnamed hypsilophodont, but it is too fragmentary to identify more precisely. It may also have belonged to a much older individual of a species similarly sized as the others in youth, for the Riches suggest that like modern reptiles (or kangaroos), dinosaurs may well have continued growing all their lives.

Leaellynasaura *[John Sibbick]*

As for the intelligence of the little dinosaurs, that is beyond our knowing. But their brains were indeed large. Judging just how large-brained they were in comparison to other animals is a task complicated by the fragmentary nature of the skulls, and the fact that they were adolescent, not adult, dinosaurs. The relationship between brain size and intelligence is also uncertain. Nonetheless, *Leaellynasaura*'s large brain with prominent optic lobes for analyzing visual information, and its peculiarly large orbits—skull sockets to hold big eyes—suggest to the Riches that these little animals "may have been best adapted among hypsilophodonts for coping with the long periods of darkness or dusk associated with polar habitats."

But how cold and how dark was it? To arrive at temperature estimates, oxygen-isotope ratios were measured by R. T. Gregory and others at Monash University in cemented lumps of the mineral calcite in the sandstone of the formation. These concretions record the amount of oxygen isotope 18 in the local groundwater, itself an average of the local precipitation in dinosaur times. The mean annual temperature was close to freezing, according to these estimates. Therefore, the winters in this humid, seasonal environment were indeed extreme. So extreme, studies indicate, that Koonwarra, a southeastern Australian site of considerable plant and fish fossils, was a cool shallow lake, ice-covered in winter, resulting in a "winterkill" of its fish. The presence at Koonwarra of cantharid beetles and mayflies, adapted to cool weather, also suggests it was seasonally cold. And rings on fossil trees indicate distinct seasons, whether changes in light or precipitation caused the varying growth rates.

It was not just cold, it was also dark in the Australian dinosaur winter, say the Riches. Australia, not yet separated from Antarctica, was well within the antarctic circle. These animals were too far from more northerly lands offering year-round daylight to have migrated to them. Dinosaurs didn't just die at Dinosaur Cove, they reared their young there.

The abundance of not just small but juvenile remains at Dinosaur Cove—more than half of the bones belong to juveniles—indicates to the Riches that this was a dinosaur nursery approximately 110 to 105 million years ago. (The dates were arrived at by comparison of spores

A well-preserved footprint of a small dinosaur from Dinosaur Cove, Australia. [Photo courtesy Earthwatch]

and pollen to other samples from eastern Australia). But the fact that young were raised at the site does not, the Riches caution, necessarily indicate that they overwintered there.

And though others have suggested dinosaurs would have to have been warm-blooded to survive long polar nights, the Riches again hold off on a verdict. "The discussion concerning whether dinosaurs were warm-blooded or cold-blooded is not sufficiently resolved," they write.

So, with scholarly caution, the Riches reason that in polar lands, perhaps some dinosaurs and other vertebrates migrated, others hibernated and became dormant, and still others may have been year-round active residents. But given the small size of the dinosaurs and the long distances those animals would have had to travel to reach winter light, it is far more difficult for the Riches or any of us to envision them as migrants. More likely they were peering cautiously and nibbling often through the long south-Australian night, or taking refuge in caves and other natural shelters.

The Australian dinosaurs seem to have evolved separately from those elsewhere in the Early Cretaceous. That, as much as their still-uncertain means of adaptation to polar life, makes them odd. They belong to a singular fauna, the only one known from dinosaur days to be dominated by little hypsilophodonts. These herbivores, like the fish fauna, are peculiar to the region, endemics as they are known to biologists. Others, the undersized allosaur that preyed upon them, the amphibians scurrying by their feet, and some of the plants, are holdovers from earlier occurrences elsewhere—relic species. They are crossovers also, having come to Australia by way of Antarctica, from more distant lands, and remained little changed in isolation.

As it is today, Australia was then a land of "living fossils," isolated then not as an island, but by its relative remoteness, or some other sort of barrier, perhaps its climate. In Early Cretaceous Australia, the small, smart, and strange dinosaurs survived.

The time of the Australian dinosaurs, the Early Cretaceous, is one of the dinosaur times least well represented in the fossil record, though Early Cretaceous dinosaurs, in Europe, were the first dinosaurs ever identified. Dinosaur-rich rock exposures dating to Early Cretaceous times have been hard to come by, but that is changing fast. In August of 1988, Thad Williams of Mill Sap, Texas, all of eight years old, was walking along a creek bed when he found the skull of a large dinosaur. Paleontologists Dale Winkler and Louis Jacobs of Southern Methodist University in Dallas supervised the ensuing dig. They excavated most of three iguanodont plant-eaters, more than fifteen feet long and a ton each.

The SMU paleontologists have also investigated another amateur's recent find at Proctor Lake in Texas. In one little depression, a dozen nearly complete skeletons of adolescent hypsilophodonts were found. Perhaps Proctor Lake was a breeding site, like Dinosaur Cove in Australia (and another long known from England), revisited year after year by hypsilophodont families. According to Jacobs, adults and young appear to have inhabited the site for hundreds or thousands of years. (Winkler and Jacobs have also excavated Early Cretaceous dinosaurs in Malawi, Africa, where huge titanosaur sauropods were the dominant plant-eaters.)

Both the iguanodonts and the hypsilophodonts evolved somewhere, sometime in the Middle to Late Jurassic. They likely found their way in those times to most every corner of a then-united world, for both are known from widely separated chunks of Early Cretaceous terrain, Australia included (the Australian iguanodon is a peculiar flamingo-snouted creature called *Muttaburrasaurus*).

Hypsilophodontids had overlapping teeth, which create a long cutting surface. Their powerful jaw muscles allowed them to grind up tough forage. This anatomy represents a major advance over earlier small dinosaur plant-eaters. Perhaps, some paleontologists think, the hypsilophodontids prospered in the Early Cretaceous because their chewing abilities made them better suited to survival in a changing plant world.

The iguanodonts, first unearthed in England in 1807, were solidly built foragers thirty feet long. They could move on two legs or all fours, and they sported large thumb spikes they may have used for defense. Iguanodont species may come in two distinct forms. That inference is based upon the historic 1878 discovery by coal miners of thirty-nine complete or partial skeletons of *Iguanodon* in a clay-filled fissure in southwestern Belgium. One is a stoutly built two-ton animal (*Iguano-*

don bernissartensis) and the other far smaller and more lightly built (*Iguanodon mantelli*), less than twenty feet long and under a ton. Both appear to be adults. Rather than two species, these may be the two sexes of a single species.

As for the dinosaurs that preyed upon these plant-eaters, not a single reasonably complete skeleton of a carnivore was known from anywhere in the world in the Early Cretaceous until recently.

In 1964 in a later Lower Cretaceous formation of Montana, John Ostrom came upon parts of his now famous *Deinonychus,* the man-size, killer-clawed hunter. An odder hunter of the Early Cretaceous was unearthed in England in 1983 by William Walker, an amateur fossil collector. In a brick quarry, Walker found the huge claw of what proved to be a thirty-foot-long, two-ton predator with a crocodile snout, now named *Baryonyx.*

What little we do know about all dinosaurs from this time suggests that they were developing a host of new forms peculiar to their habitats. This is in keeping with evolutionary theory, as across the world in the Early Cretaceous continents were separating, and new dinosaurs evolving in restricted ranges.

Familiar forms do persist, like stegosaurs in Africa and India, and the squat armored nodosaurs lately found in Early Cretaceous Colorado rock. But from Africa in the Early Cretaceous come giant spine-backed predators and plant-eaters. In Asia, small predators and the most primitive ancestors of horned dinosaurs evolve. Sauropods decline in many places, but in South America, they develop armored sides and club tails, and others sport twin rows of spines down their backs.

But there is another factor aside from geography that determined what triumphed and what died out in the Early Cretaceous. Albeit slowly and far from uniformly, flowering plants took root then (the first butterflies appeared at the same time). These plants are not well known from the first 15 million years of the period, when the flora looks much as it did in the Late Jurassic. But by 120 million years ago, flowering plants in the equatorial regions were diversifying, even as the vegetation in lands nearer the poles—from Europe, the American West, and Siberia in the northern hemisphere, to South Africa and Australia in the south—was still largely fern- and conifer-dominated forests, with an increase in ginkgoes, not flowering trees.

By the Late Cretaceous, flowering plants were well distributed around the world, though they did not achieve a status approaching their modern diversity and range until the last 15 million years of the dinosaurs' lives. Flowering plants did not bloom into complex rain forests until after dinosaur days. But flowers were likely abundant in the Late Cretaceous as weedy plants in the understory of open forests, and the dominant vegetation in the open woodlands. The hypsilophodonts, whose kind endure until near the end of dinosaur days, may have had the dental anatomy to crop the new flowering plants efficiently. But the iguanodonts disappear in the Early Cretaceous, to be replaced by a proliferation of duckbilled dinosaurs each equipped with a prodigious battery of grinding teeth and flexible jaws well suited for digesting all manner of plants.

Gone to Extremes

In the search for more information on the plants of the Far North at the end of the Age of Dinosaurs, more clues were added to the mysteries of dinosaur adaptation to polar life.

Today, coastal Alaska is cool, wet, and well vegetated. In winter the temperature drops below freezing, but not significantly so. Summers are moist and temperate. Shrubs and trees abound along the coast, and along rivers, lakes, and in swamps. Some of the continent's largest animals feed upon the plants and animals that thrive in the long summer daylight. So conducive are the long days to plant growth that in prize-winning Alaskan gardens, cabbages grow to the size of ottomans. In wilder parts of southeast Alaska, bears, mountain goats, and Sitka deer prosper on the abundant flora.

But this bloom of life pales by contrast to life in the Late Cretaceous in another part of the state. About 70 million years ago, some 5 million years before the end of the dinosaurs, the North Slope of Alaska was such an Eden, at least in daylight months. Plants and trees that had shed their leaves and needles the previous fall grew them back with a vengeance come spring. Vines ran riot, carpeting the landscape.

A singularly rich plant fossil record, including tree rings from seasonal growth, supports this scenario. There is, according to the paleobotanists who've rafted up and down the rivers of the North Slope, no place else on earth with such a rich plant fossil record—whole forests, with wood from dinosaur times preserved so nearly intact that it still burns. Animals would have traveled far, as do caribou and musk-oxen for far less these days, to browse in the luxury of these forests.

Throughout the late 1980s, one team of scientists studied the fossil remains of this rich world. Each August they journeyed to the banks of the milky Colville River in far northern Alaska. They reached the fossil site by chartered aircraft from Prudhoe Bay, to a sandbar on the river, then slowly, on an outboard-powered rubber raft, impossibly weighed down by the crew of six. Their site was accessible by foot as well, if one clambered up the hundred-foot-high and treacherously loose, brown siltstone cliffs and along the overbank. That route leads also to close encounters with obliviously grazing caribou, to the flushing of frightened ptarmigan, and to openmouthed observations of the aerial acrobatics of young peregrine falcons chased by rough-legged hawks. The team had only weeks to work in this wilderness. In July, the endangered peregrines were nesting and could not, by law, be disturbed on these federal lands. By mid-August snows had begun obscuring the site.

The scenery is transformed in winter. The river freezes and ice dams form upstream. In spring the dams break and a wall of water scours the Colville River valley cliffs. Each spring of the excavation years, this torrent washed away any traces of the previous year's digs, fossils included. But the scouring exposed anew just as many other bones. If researchers saw something they wanted here, they had to get it out of the ground, into plaster, and onto the airplane in two weeks, or it was likely to be lost forever.

In the summers of 1987 through 1989, the fieldwork at Ocean Point, as this site is known, was under the direction of the chairman of the paleontology department at the University of California at Berkeley, William A. Clemens, an erudite, squarely built, and gray-bearded man

On the slopes of the Colville River in Arctic Alaska, University of California/Berkeley mammal paleontologist William Clemens digs for dinosaurs.

nearing sixty with an avuncular manner, bushy eyebrows, and resounding voice reminiscent of Walter Cronkite's.

Until recently Clemens was perhaps best known for his studies of North American mammals and their evolution across the boundary that marks the end of dinosaur time, in the place where that change is best recorded, Hell Creek, Montana. In the past decade, Clemens has emerged as one of the leading spokespersons for the gradualist perspective on dinosaur extinction. Since coming to the Colville, Clemens has plunged into the business of finding hundreds of dinosaur bones from among the last of the dinosaurs.

The discovery of the first dinosaur remains from the North Slope of Alaska, was, like many dinosaur discoveries, entirely serendipitous. The chances of stumbling on dinosaur fossils are few indeed in northern Alaska. There simply aren't many places likely to show fossils in the flat and thickly carpeted tundra. Where fossil-bearing formations are exposed is along the high and fast-eroding banks of North Slope rivers, rivers such as the Colville.

Clemens learned of the promise of the Colville fossil beds in the early 1980s from Rip Repenning, a United States Geological Survey scientist based in Menlo, California. Repenning sent Colville dinosaur fossils, collected by USGS staffers, to Clemens. The fossils intrigued Clemens. He drafted a proposal to USGS to look for them in the banks of the Colville, fossils that might indirectly reflect on the actual age of the sediments. In 1985, Clemens had the funding to do fossil research.

Among the first dinosaur bones Clemens's crew came upon in their digs were the teeth of a *Troödon*, the smart little predatory dinosaur known from Asia and North America. And on one edge of the sandbar by the camp, while team member Howard Hutchinson was screen-washing shovels of sediment in search of small fossils, he spotted a chip from the snout of a horned dinosaur in the gravel by his feet. He dug into the cliff behind him and found most of an intact skull of a horned dinosaur known from the North American West, *Pachyrhinosaurus*. The vast majority of the bones that protrude from and cluster within the cliff-face are broken bits of duckbills, not horned dinosaurs—ribs, vertebrae, femurs by the scores. Most appear to be from a species previously known from Saskatchewan.

The close resemblance of the dinosaurs found along the Colville to those known from thousands of miles away on the North American continent at the same time suggests to many scientists that the arctic dinosaurs were summer visitors to this sometimes inviting, sometimes inhospitable land. But not to Clemens. The principal residence of the animals that died along the Colville River at the end of dinosaur days is an issue fraught with repercussions, not only for dinosaur behavior, but dinosaur physiology and extinction as well.

Clemens and his collaborators suggested that the juvenile animals whose remains were frequently found at the site would have been too small to migrate in winter. Nor did the animals need to migrate, as they could survive a cool but nonfreezing winter. They could find winter food by straining the mat of vegetation under the water surface or munching on woody ground cover or evergreen conifers.

Clemens believes that the year-round residence of dinosaurs in the cold and the dark is a refutation of the supposed "lethal effects" of an asteroid impact hypothesized by Alvarez and others as the cause of dinosaur extinction near the time of these arctic dinosaurs (see Chapter 11). Those effects, in Clemens's words, would have included "a period of darkness lasting a few weeks or months and a great decrease in ambient temperature." If the arctic dinosaurs did not migrate, says Clemens, this "provides direct evidence of the ability of some species to tolerate up to several months of darkness. Thus, some of the proposed effects of impacts of an asteroid or comets, increased volcanism, or related hypotheses may not have been the direct cause of the demise of dinosaurs."

Clemens adheres to that view to this day. He notes that his crew has found many small dinosaur teeth at their site. These may be the teeth of small adult dinosaurs, better equipped by their small size to reside in arctic climes, like the little hypsilophodonts of Australia. Or they may be babies, suggesting that the Arctic was also a dinosaur nursery, and so likely a year-round home, for to grow large enough to migrate within a few months' time seems a daunting, if not impossible, challenge. As yet, no one has found solid evidence for an arctic dinosaur nursery—eggs or nests. But many environments that preserve bone are too acidic to preserve eggshell.

How would resident dinosaurs, young and adult, have coped with the cold and dark? Clemens suggests they could have sharply reduced their food intake and switched into a lower metabolic rate, much as bears do today. Again, the evidence for this, perhaps from growth rings in dinosaur bone, has yet to be obtained.

Whether dinosaurs migrated from Alaska in the coldest, darkest months is unknown, but the subject of dinosaur migration has been much studied. The famous German paleontologist Von Huene introduced the concept of dinosaur migration in 1928, postulating their movement from dry to moist environs. The notion of dinosaurs migrating north and south seasonally was raised by Canadian paleontologist Dale Russell in 1973 and discussed at greatest length by Smithsonian paleontologist Nick Hotton.

As Clemens had tied his concept of resident arctic dinosaurs to a refutation of a catastrophic-extinction theory, so Hotton linked his ideas of dinosaur migration to a dissenting view on dinosaur warm-bloodedness. His arguments appeared most extensively in his 1980 contribution to *A Cold Look at the Warm-Blooded Dinosaurs,* entitled "An Alternative to Dinosaur Endothermy: The Happy Wanderers."

Hotton reasoned, "All large dinosaurs, herbivores and predators alike, would have had to be wanderers in order to gather enough food to keep themselves going." By moving, slowly and steadily, dinosaurs could have warmed themselves without being warm-blooded. By moving north or south in keeping with moderating temperatures, they would remain away from climatic extremes that would chill or overheat them. Dinosaur migration could thus be compared to the sun-following movements of lizards, scaled up to giant size. Without any apparent insulation, dinosaurs, according to Hotton, could not have survived in latitudes higher than sixty degrees in winter. So, "the fossils at those latitudes reflect summer occupancy only."

Why, then, bother heading north at all? Those who have seen the giant cabbages at the Matanuska, Alaska, fair needn't ask. Twenty hours of sunlight a day can do wonders for vegetation, and for satisfying the appetite of a hungry herbivorous dinosaur.

Cued by changing temperature or daylight length, the animals would begin their migration. Hotton calculated that no dinosaur would have had to journey more than two thousand miles each way, the distance required to traverse thirty degrees of latitude. "With four to six months to complete the trip each way, they would need to have averaged from twelve to seventeen miles per day." In his projections, Hotton walked dinosaurs just half a day, at a speed half those of the slowest sauropods, as measured from their stride length and size in trackways. He even took into account that the mud flats where those footprints were recorded probably permitted faster travel than overvegetated dinosaur migration routes.

Herding might explain the trackway "highways" of thousands of footprints of iguanodonts found in Early Cretaceous rocks from Colorado to northern New Mexico by University of Colorado paleontologist Martin Lockley. Herding would also explain the fossil discoveries in a single locale of groups of dinosaur competitors, animals that would likely have vied for a common food source in life. For these animals to share a habitat would be a violation of an axiom of ecological theory—the partitioning of resources. But where two or more different kinds of hadrosaurs are found in abundance, as at Dinosaur Provincial Park, they may not have been directly vying for the same forage in the park. Rather, both may have been passing through, at varying times, in their respective herds.

The pattern of distribution of some dinosaur fossils suggests a herding route to Phil Currie, a strong supporter of the theory that Alaskan dinosaurs were summer migrants. The diminished presence of Late Cretaceous fossils south of Montana suggests to him that "the region was at the southern limit of some migration routes." The same is not true to the north. A bone bed of the same genus of horned dinosaur found in Alaska, *Pachyrhinosaurus,* was excavated in southern Alberta in 1945. Another was excavated by the Royal Tyrrell Museum's crew 450 miles farther north in Alberta in 1986. Add to these the skull found by Hutchinson along the Colville the next year, two thousand miles north of the southern Alberta bone bed. Either that animal was adapted to live year-round in each of these distinct habitats, or it commuted between all. Whatever their strategies, Cretaceous dinosaurs from Alaska to Australia were well equipped for dealing with an increasingly challenging world.

Fossils into Dinosaurs

There is something remarkable about the dinosaur fossils of Alaska, a feature that offers the prospect straight from Michael Crichton's *Jurassic Park* (or vice versa) that dinosaur tissue may one day be recreated. To sense their uniqueness, you need only pick one up. While most fossils have a rock's heft, these are, as Phil Currie says, "as light as dog bones a few days old."

British fossil plant expert Robert Spicer, among the first to gather these fossils, speculated that their lightness owed to the fact that they were not buried deeply and so underwent no great heating and pressure. Rather, they were on ice. "They were kept cold a long time, slowing down their alteration."

Spicer sent several of the Alaskan fossils to a geologist acquaintance at the University of Glasgow, Gordon Curry. Since 1984, Curry and his collaborators have sought to ascertain the structure of the organic compounds that occur in rocks. Far from devoid of life, rocks contain the remains of living tissue—ten thousand times the organic compounds of life on earth today.

Until recently, the structure of these compounds had been largely a mystery. Death, decay, and fossilization were long thought to destroy the amino acid chains that are the building blocks of protein, and of life. But as Curry has observed, some of the more resistant organic molecules can be incorporated into rocks with only minor changes. These biomarkers have broken down into their constituent amino acids, but are still sufficiently preserved to enable scientists to track molecules to their origins.

The sequence of those amino acid molecules differs from individual to individual as well as from species to species.

DNA holds the genetic code, but it is locked in soft-tissue cells, seldom preserved in recognizable fossils. (Crichton solved the problem by having his fictitious scientist locate DNA from tissues preserved in fossil amber, a possibility explored since 1982 by paleontologist George O. Poinar of the University of California/Berkeley.)

DNA may be hard to come by in fossils, but amino acids can now be detected in fossils from almost incomprehensibly small amounts. Challenged by an appropriate antibody, fragmented proteins produce a detectable chemical reaction. W. Dale Spall, a geochemist at Los Alamos National Laboratory, asserted in 1991 that he had isolated proteins from a backbone of *Seismosaurus*. But Spall acknowledged that the proteins may have seeped into the fossil from other sources, and colleagues have also questioned the origin of the proteins. To stir up the proteins in the Alaskan fossils, Curry is experimenting with osteocalcin, a monoclonal antibody shown to be effective on fossils up to 70 million years old (perhaps 5 million years older than the Alaskan fossils).

Curry believes that familiarity with Alaskan dinosaur fossil proteins could allow him to identify even minuscule fossil fragments down to the appropriate group, even to sort out evolutionary pathways according to similarities in molecular structures between different dinosaur groups. By such methods, he believes, "we might also tie down the rate of evolution, and build up data on how quickly dinosaurs were changing per million years."

And perhaps, given their deep protective freeze, the relatively mineral-free Alaskan dinosaur bones have preserved some of their soft-tissue DNA. Says Curry, "This is totally speculative, but with the sequence of proteins in their DNA you could regenerate a piece of dinosaur, a very small part of a molecule, but a dinosaur's nonetheless." A laboratory technique invented in 1985 called polymerase chain reaction allows researchers to multiply, many times over, tiny amounts of DNA, even fragments of ancient molecules. One day, then, it may be possible to replicate a dinosaur's DNA from an airy chunk of an Alaskan duckbill's bone.

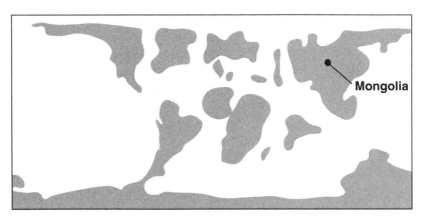

The breakup of the supercontinents, 83–65m. yrs. ago

T	J	C

83-65m. yrs. ago

245m.   65m.

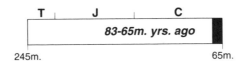

AGE OF THE DINOSAURS

Triassic Jurassic Cretaceous

First animals		First land animals	T	J	C		First human

590m.   350m.   245m. 208m.   145m.   65m.   Today

The Gobi Dinosaur Chase

In the broiling heat of a July midafternoon in 1988, Dong Zhiming was on his hands and knees, scraping at the white-crusted Mongolian Gobi, when he came upon a flattened black shard nearly as large as his palm. Brushing clean the object with his thick fingers, Dong chortled with glee.

Dong and more than two dozen Chinese and Canadian co-workers had come to this remote wasteland on the most costly of all dinosaur explorations. Their mission was to find fossil evidence of the relationship between the two best known of all dinosaur populations, the Late Cretaceous dinosaurs of Mongolia and of North America. But what Dong had found was no fossil.

Holding the find in his outstretched hand, Dong rushed across the plain to show it to his colleague Phil Currie, who was hunched over a mound of scattered pebbles and bone fragments. Currie held the piece against the glaring sun and shook his head, chuckling. There was no mistaking the find, nor its origin—a crushed tin, etched with the outline of an American family in its Dodge convertible. Below the auto, bold letters proclaimed SPIRITS.

This particular piece of American litter was a special delight to Currie, for it belonged to the hero of his childhood: Roy Chapman Andrews.

"Imagine some explorers coming along a million years from now and digging us up petrified in a touring car." So Roy Chapman Andrews commented, with characteristic good cheer, to his fellow dinosaur-finders on the American Museum of Natural History's Central Asiatic Expedition of the 1920s. As they were in the midst of the Mongolian Gobi, battling armed brigands, the political uncertainties of a land torn by civil war, 140-degree heat, week-long sandstorms, and infestations of vipers, what Andrews blithely envisaged must not have seemed an unlikely fate to his companions.

The Dodge Motor Company had presented tin flasks to members of the Andrews expedition. Here now—sixty-odd years later—was one of those containers—proof that Currie, Dong, and fellow members of the joint Chinese-Canadian Dinosaur Project had found their way to the precise locale of the Iren Dabasu ("Shining Telegraph Station") quarry. This was the site of the first dinosaur bones ever found in central Asia.

Since the 1920s thousands more bones have been found in Mongolia, by Andrews and by Chinese, Swedes, Soviets, Poles, and Mongolians after him. The discoveries of nests, intact eggs, footprints, and skeletons of tens of familiar and unfamiliar dinosaurs from the Cretaceous have made the Gobi one of the most productive of all dinosaur-hunting locales. Yet the dinosaurs of Mongolia and their habits and habitats remained one of the great dinosaur mysteries. Now, more than sixty years after the Andrews expedition first attempted to answer many of the questions about these dinosaurs, the Canadians and Chinese had come to the same ground to reexamine the evidence and question the earlier conclusions.

Locating anything is difficult in the Gobi, a huge and shifting arc of land, some one thousand miles long and six hundred miles wide, straddling central Asia along the border between Inner Mongolia (now the Nei Mongol Autonomous Region of China) and Outer Mongolia (now Mongolia, until recently the Mongolian People's Republic). It is a forbidding desert. In the Mongol language *gobi* means just that—"waterless place." The Chinese refer to it as a "desert of sand and gravel." To geologists, this and other gobis are small, open, level-surfaced basins within broader basins.

Most of the Gobi is a gray-brown elevated plain, carpeted with pebbles and in some places, dinosaur bones, over a rock floor as hard and corrugated as a washboard. Around springs and after rainfalls, brown patches of tamarisk and saltwort green up, and the Gobi takes on the look of the steppe pastureland that surrounds it.

Even if it does not conform to the sand dune expanses we imagine deserts to be, the Gobi is decidedly dry country, drier even than in Genghis Khan's day. He and his armies rode through grassland, even forest, in what is now the dusty Gobi. The temperatures today are cruel—a swing of two hundred degrees between summer and winter extremes—but the harshest season may be the spring, when roaring winds blow one sandstorm after another across the plain and south through China. Beijing often turns orange in April from Gobi dust. The silt-forming grains that blow south from the Gobi are known as loess (German for "loose"). Gobi loess is what yellows the Yellow River and its fertile valleys. It has accumulated to depths of a thousand feet where it commonly falls in northern China.

Each year erosion exposes more dinosaur bones, but also obscures the way to them—winding, unmarked caravan routes, centuries old, which are the only roads through the Gobi, save an arrow-straight paved military highway to the border.

That highway opens into the Inner Mongolian border town of Eren-hot, a drab outpost of ten thousand lonely souls scattered about wide, empty avenues. And just miles outside town, on a patch of the endless pebble and scrub plain marked by an almost-dried white soda lake, lie the ruins of a still-functioning soda mine, and the low mounds left decades ago by bulldozing Soviet paleontologists.

Here, the Chinese and Canadian paleontologists don't expect to find any spectacular specimens. Ancient stream channels appear to have cut and swept the bones into small fragments, fractured and concentrated further by the mound-building Russian explorers. Yet Dong has already located a bed of broken, jumbled blue-gray hadrosaur bones, which a team of provincial workers is busily unearthing for a regional dinosaur museum still in the planning phase. Currie has come across the teeth and foot bones of several predatory dinosaurs, some common, others rare, and all known from both Asia and North America. The presence of all these dinosaurs, even in such fragmentary form, is compelling evidence that Asia and North America shared their dinosaurs late in the dinosaur era, a subject of considerable controversy and confusion before and since Andrews came to the Gobi.

Andrews's Grail

On April 27, 1922, in Iren Dabasu, Inner Mongolia, very near where Currie and Dong now worked, American Museum paleontologist Walter Granger walked over to his colleague, eyes shining, his pocket full of fossils, and announced, "Well, Roy, we've done it. The stuff is here." At least that's how Andrews remembered the first dinosaur discovery from Central Asia, in his several popular accounts.

This is the stuff of legends, and so is the life of Andrews, in truth and in his own colorfully embroidered accounts. To the Chinese and North American paleontologists who've trod in Andrews's footsteps, he was no hero. He tricked his hosts, misidentified discoveries, allowed racist notions to influence his science, and so butchered the fossils he excavated that to this day, museum workers say a poorly prepared fossil has been "RCAed." But to millions of Americans in the early decades of the twentieth century, Andrews was a hero of epic proportions. And his globe-trotting adventures have produced a rich scientific legacy.

As a nature-loving Wisconsin youth, Andrews wanted nothing more than to work for the American Museum of Natural History. He wangled a job sweeping floors and working in the museum's taxidermy department when he was fresh out of college and didn't leave the museum's employ for decades. Within a few years of joining the Museum, he'd explored the wild forests of Korea, searched for a reported blue tiger in southern China, hunted whales off the coasts of Alaska and Japan, and spied for the Allies in China in World War I. And in 1921, Andrews convinced Museum president Henry Fairfield Osborn to authorize him to lead a motorized expedition across the Gobi.

Andrews was pushing a long shot. He had seen promising-looking outcrops for fossils while hunting for game in the Gobi. But the Gobi was a little-explored wilderness that had turned out, to his day, but one fossil—the tooth of an ancient rhinoceros, widely believed to have been dropped there by passing traders. Previous explorers had made only halting inroads by horse and camel, but Andrews had tried the Gobi by motorcar in his wartime espionage and knew he could traverse great stretches if properly equipped.

Previous explorers' dim results notwithstanding, Andrews had little trouble convincing Osborn that the Gobi would be an excellent spot to test one of Osborn's pet theories—that Asia was the wellspring of human, and other animal, life. In Andrews's day, before the movements of the earth's plates were known, ancient land bridges, now submerged, were thought to have once united continents, permitting animal migrations.

Andrews sought to demonstrate that man had traversed the Asian-American bridge, and spread in other directions as well, many hundreds of thousands of years before. The notion, as Stephen Jay Gould has pointed out, was fundamentally racist. Asia would have been preferred to black Africa, the likelier root of human life. Perhaps in belated awareness of the bias of Osborn's theory, or at least his lack of success in making a case for it, Andrews retrospectively downplayed the human-origins angle of the expedition. (The press did just the opposite, even before Andrews set out, referring to the trek as "the Missing Link Expedition.")

Just six days before they found the first dinosaur remains at Iren Dabasu, Andrews's team had set out north from Beijing on what he'd call the "new conquest of Central Asia" (the title of his seminal work) with a force of forty men, five vehicles, and seventy-eight camels. They'd headed to Iren Dabasu, attracted by what Andrews had heard from Mongols was there: " 'Bones as big as a man's body,' they said."

At Iren Dabasu and elsewhere in the Gobi, Andrews and his crew found plenty of fossil treasures if not evidence of an Asian ancestor to modern man: they retrieved artifacts from Stone Age humans, skulls of tiny Cretaceous mammals and huge postdinosaurian mammals such as the twenty-foot-high *Baluchitherium,* purported to be the largest mammal ever to walk the earth (it turned out to be identical with *Indricotherium,* an earlier find from India). They also found instantly recognizable bits of titanotheres, rhinoceroslike early mammals—the first of their kind discovered outside America.

Within a year of traipsing farther into Mongolia, Andrews's team found lots of dinosaurs. Most common was a small herbivore, scarcely six feet long in adulthood. With its narrow beak and well-developed neck frill, this creature appeared to be ancestral to the ceratopsians, even though it did not have horns. Ceratopsians were the horned dinosaurs, including *Triceratops,* already widely known in Andrews's day to have once lived in the North American West. The newfound frilled and hornless dinosaurs were named *Protoceratops andrewsi* ("Andrews's first horned face"). Andrews saw these ceratopsians, like the tiny mammal skulls, as evidence that Central Asia had been the point of origin of North American Late Cretaceous animal life.

The most fortuitous finds of the Andrews expedition came some two hundred miles to the northwest of Iren Dabasu. In the midst of the hard, flat Gobi, Andrews had come upon a magnificent red sandstone canyonland, "paved with white fossil bones," that seemed "to shoot out tongues of fire" in the late-afternoon sun. He called the spot the Flaming Cliffs.

Here were *Protoceratops* in abundance. And as team paleontologist George Olsen discovered, the clifftop held a nest of crushed, oblong, nine-inch-long eggs. Atop one clutch Olsen found a strange toothless predatory dinosaur, which Osborn dubbed *Oviraptor* ("egg thief"). Andrews speculated that the eggs, and others nearby, twenty-three in all, were from a *Protoceratops.* He announced them to be the first dinosaur eggs ever found.

They weren't. French researchers had found eggshells near the Riviera years before, and Andrews himself had done so just a year previously, but as there were no other dinosaur fossils near the finds, they weren't correctly identified. Nor were Andrews's eggs necessarily *Protoceratops.* That dinosaur was just the closest likely candidate because of its abundant nearby remains. Nonetheless, the Flaming Cliffs egg-find, above all his discoveries, brought Andrews celebrity beyond even his fond expectations. The notion of egg-laying, nest-sitting dinosaurs captured the public's imagination.

Andrews made the most of that phenomenal interest. He sold one egg for $5,000 at auction to Colgate University to benefit the American Museum, an act that outraged Chinese authorities. If one egg from China was worth that much, they reckoned, how much was Andrews going to make from their fossil treasures?

By the end of the decade, the economic woes of the American Museum, its benefactors, and world and Chinese internecine political turmoil compounded Andrews's diplomatic difficulties. Andrews met with President Hoover in an unsuccessful attempt to coerce cooperation from the Chinese. J. P. Morgan helped win him another year's grace in 1929 by threatening to withhold support for the Nationalist military government in China. Still, Andrews gave up the Gobi, reluctantly, with his 1930 digs in Inner Mongolia.

After Andrews

But where Andrews's expedition ended, others followed. And with each expedition's dinosaur discoveries the prevailing scientific wisdom about whether dinosaurs moved between North America and Asia changed.

Sven Hedin, Swedish explorer and geographer, and later Nazi apologist, undertook his own survey of Central Asia in 1927, with German support. Though Hedin's health failed and an Asian ancestor to modern man eluded him, the joint Swedish-Chinese investigations stretched into the 1930s and turned up more dinosaurs, among them a bulky new sauropod. One site noted by the Sino-Swedish team as a outcrop of sandstone formation similar to the Flaming Cliffs but two hundred miles to the southeast was marked on Hedin's maps as Ulan Tsonchi, named for its landmark red tower. Hedin's party found only scraps of dinosaurs there, though its environs would, ultimately, surpass the Flaming Cliffs as a dinosaur treasury.

From 1946 through 1949, Soviet paleontologists carried out extensive digs in the Gobi, unearthing several well-preserved large-dinosaur skeletons. Among them was a giant predator astonishingly like the North American behemoth *Tyrannosaurus rex,* and one of the largest of the duckbills, belonging to a genus also known from North America. Both finds were strong evidence that considerable interchange between Asian and American dinosaurs occurred from 80 million to 65 million years ago in the Late Cretaceous. The explorers also came upon the monstrous claws of *Therizinosaurus* ("scythe reptile") with lethal talons nearly two and one-half feet long. What this huge animal looked like is still anyone's guess.

A return to Iren Dabasu was not attempted until Chinese and Soviet paleontologists came back to the Gobi in 1959 and 1960, digging up Andrews's sites with a bulldozer. They found no new species there, and the expedition disbanded in political acrimony.

The explorer's gauntlet thrown down by Andrews was next picked up by a Polish woman scientist, Zofia Kielan-Jaworowska. Interested in Cretaceous mammals, the headstrong then-director of the Paleobiological Institute of the Polish Academy of Sciences (now a researcher in Oslo, Norway) organized four of her own expeditions to the Gobi between 1963 and 1971, with Mongolian cooperation.

Kielan-Jaworowska was not deterred by any of the daunting logistical obstacles to Gobi exploration. And when her Mongolian guide succumbed to too much Soviet "bozhomoika" vodka (literally, "Oh my God!") she drove one of the huge expedition trucks herself.

The Polish expeditions were hugely successful. Their most spectacular discovery still ranks as the most dramatic of all dinosaur fossil finds. Thirty kilometers west of the Flaming Cliffs, the Poles uncovered two dinosaur skeletons beautifully preserved just as they may have died, locked in a vicious struggle. A swift small predator, *Velociraptor,* holds the skull of a *Protoceratops andrewsi* between its forelimbs, hind claws locked in its victim's abdomen. The *Protoceratops*'s jaw is clamped on the *Velociraptor*'s hand. (The "Fighting Dinosaurs" are on view at the State Museum in Ulan Bator, Mongolia, a stirring reminder that fossilization can preserve animals as they appeared in the very moment of death).

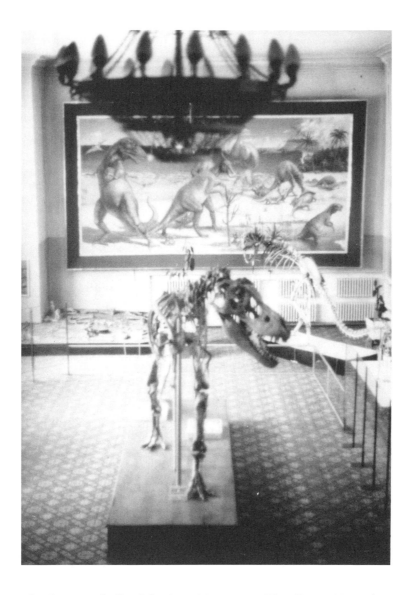

The dinosaur halls of the State Museum in Ulan Bator, Mongolia, house some of the most spectacular and least known dinosaurs in the world. In the foreground is Tarbosaurus, *a close cousin of* Tyrannosaurus rex. *[Photo by author]*

Kielan-Jaworowska's team attempted, in a way those before them had not, to interpret the manner in which these animals lived, died, and evolved. Since Polish sedimentologists, the Russians, and to a lesser extent Andrews's team before them had interpreted the formation of the Flaming Cliffs as lakeshore environment, Jaworowska speculated that the opponents "died as they fought . . . embedded in quicksand." A nearby find of a *Protoceratops* skeleton, propped upright in the sand, further suggested to the Polish researchers that the animal died trying to escape from the quicksand.

To this drama, Kielan-Jaworowska added solid scientific evidence to refute Andrews. Her team collected 150 mammal skulls from the Flaming Cliffs, perhaps 80 million years old, and from younger sediments as well. The peculiarities of these specimens, together with other contemporary discoveries of mammals in North America, showed that Andrews's notions about the origins of placental mammals in the Gobi were oversimplified.

In the southwest Gobi, Kielan-Jaworowska found two younger formations than that of the Flaming Cliffs, and more evidence Andrews may have been off-line on his ideas of dinosaur origins. While only fragments of sauropods, the herbivorous giant dinosaurs, had been known from the Gobi, and none from the northern United States and Canada in the Late Cretaceous, the Poles found a fine skull of a new sauropod. Stranger still were the huge forelimbs and shoulder girdle of a puzzling giant the Poles named *Deinocheirus.* Here was a carnivore the apparent size of *Tyrannosaurus rex,* some forty-five feet long. But instead of *T. rex*'s puny forelimbs, it had extremities eight feet long, fixed with foot-long claws.

Whether *Deinocheirus* was related to *Therizinosaurus,* the even more monstrously clawed Soviet find in the Gobi, remains uncertain. Nothing remotely akin to either dinosaur had ever been discovered in North America. Nor were the known dome-headed dinosaurs of North America much like the three new dome-head skulls the Poles turned up in the Gobi. And where North America was rich in horned and duckbilled dinosaurs, Asia had none of the true horned dinosaurs and few of the duckbills.

Kielan-Jaworowska concluded, largely on mammalian evidence, that there was no land connection between Asia and North America large enough to permit an interchange of large animals in the Late Cretaceous. She proposed that the tiny mammals were able to cross over, suggesting that they and a few other animals made it across the continents by fluke, perhaps by fording or rafting shallow marine straits.

In 1971 the jealous Soviets put an end to further work by the Poles and reinstituted their own investigations in the Gobi. Soviet researchers and their Mongolian coworkers have consistently worked the Gobi for the last twenty years. They've uncovered dozens of sauropod and *Protoceratops* eggs, nearly complete skeletons of tarbosaurs (*Tyrannosaurus rex*'s close cousin), a baby hadrosaur, and *Protoceratops* of several sizes—far more riches than they've had time and resources to prepare and study in full. They have, however, described several new species of dinosaurs, ostrich-mimic dinosaurs and theropod predators in particular. Though rare elsewhere, carnivorous dinosaurs are strangely abundant in the Outer Mongolian Gobi—twenty of the world's fifty known forms have been discovered there. Many of those specimens are now on display in the dinosaur hall of the State Museum in Ulan Bator.

However, aside from descriptive papers in Soviet scientific journals little distributed and seldom translated in the West, the Mongolian dinosaurs are little known outside their home, despite the outstanding efforts of the Poles.

Together, independently, and occasionally with Mongolian scientists, two Polish women paleontologists from Kielan-Jaworowska's team, Halszka Osmolska and Teresa Maryanska, wrote a host of highly regarded papers in English based on their collections from Mongolia. Among the finds they described was a baby protoceratopsian dinosaur skull less than an inch long—at the time the smallest dinosaur skull ever found. Their work, however, goes well beyond mere description. They've written on the ecology of three dinosaur-bearing formations in Mongolia, and on possible salt-concentrating glands in the noses of desert-dwelling dinosaur predators. Coincidentally, they undertook a cladistic analysis of bird-hipped dinosaurs at the same time as Paul Sereno did his. Without funds or permission to return to Mongolia, they are still working in Warsaw on the material they collected in Mongolia more than two decades ago, as Osmolska says, "completing the inventory of the Late Cretaceous Mongolian dinosaurs." That inventory has lately been expanded enormously.

Tarbosaurus *[John Sibbick]*

SIBBICK

The Dinosaur Project

As a childhood admirer of Andrews and now one of the world experts on the anatomy of meat-eating dinosaurs, Phil Currie had always put a visit to Mongolia firmly at the top of his wish list of research trips. He'd been asked to compile just such a list by an idealistic young researcher for the Tyrrell Museum publications office, Brian Noble, in 1982. Noble then pursued, in vain, a cooperative project with the Soviets and Mongolians that would allow Currie to dig and study in Mongolia. Among Western dinosaur researchers, only Paul Sereno had been to Mongolia in decades.

But through its generous terms, Noble's energy, and the intercession of Canadian government and business leaders, the Ex Terra Foundation did land Currie's second choice, Inner Mongolia in China. The multiyear program would cost more than $7 million, but Canadian and Albertan government loans and receipts from an international exhibition of the project's fossil finds would, at least in theory, match the expenses.

Lately involved in many international scholarly exchanges, the Chinese had received several offers for joint dinosaur study, American, Japanese, and French among them. But Noble's proved particularly appealing because it was unique in proposing visits by Chinese paleontologists to Canada to study North American dinosaur faunas.

Gallimimus *(OPPOSITE) and segnosaur* Nanshungosaurus brevispinus *(BELOW) [John Sibbick]*

Canadians would pay for those visits and their own trips to China. After preparation in Canada and a world tour (scheduled to open in Edmonton in 1992), the fossils would be returned to China.

The Chinese added other conditions. Though Currie and Canadian project coleader Dale Russell had intended to study Late Cretaceous dinosaurs, the Chinese opted to expand the search to the Jurassic of far-western China, in Sinkiang, an area they had lately worked and one of particular interest to head Chinese dinosaur scientist Dong Zhiming, who'd made four missions to the area in search of his favorite quarry, Jurassic sauropods. Russell was also eager to study sauropods, animals most abundant in the Jurassic.

The team's most significant finds in its first year of exploration, 1986, came not from China, but Alberta, where fossil worker Tang Zhilu found the braincase of that smartest of all dinosaurs, the predatory *Troödon* (see Chapter 1).

The first full season of Chinese fieldwork was performed in the summer of 1987 in Sinkiang, three hundred miles northeast of Urumqi, the most inland city on earth, near remote Chinese missile-testing installations. Chinese officials balked at providing the crew geological or topographic maps to some of these sensitive locations.

The fossil localities lay in the midst of spectacular red- and yellow-banded badlands. The dinosaur-bearing strata were stubbornly hard red rock, so resistant to excavations that jackhammers, even dynamite blasts, were needed to remove the overburden. The brutish fossil-extraction methods of some Chinese were disturbing to the Canadians, especially when exploded fossil turtle fragments rained down on scientists eager to study animals in more intact form.

Other obstacles to paleontological progress had more to do with culture and personality than scientific method. Currie, Dong, and Russell waded into the quarries with pickaxes, though they might normally have done more prospecting, in order to provide a role model for the large Chinese crew (thirty Chinese to eleven Canadians), many of whom lacked both experience and enthusiasm. Dong and equally headstrong Canadian project leader Brian Noble feuded, while Russell, the dreamy theorist, and Currie, the pragmatic field man, were an odd pair. Canadian field-workers' fraternization with their Chinese counterparts was discouraged, and one exuberant Canadian was reprimanded for dancing too close to a Chinese woman.

Frictions notwithstanding, the Sinkiang effort produced excellent results. Researchers uncovered a tritylodont, several extinct turtles, and a rare find, the skull of a pterosaur. Twenty miles from camp they came upon a fossil grove to rival Arizona's Petrified Forest. The four-foot-high stumps of twenty-three conifers stood, their internal structure splendidly retained, among fallen trunks, one seventy-five feet long and four feet in diameter. In the midst of this now arid plain, the fossil trees recalled a Jurassic valley comparable to today's Great Salt Lake environment, with vast shallow alkaline lakes and scrubland in its

Currie's co-leaders on the most elaborate dinosaur hunt ever staged were Dong Zhiming of China and Dale Russell of Canada, seen here resting in the brutal Gobi noon sun. [Photo courtesy Ex Terra Foundation]

basin, and forests on its higher slopes that had been dominated by towering evergreens 180 feet high.

From the Sinkiang basin, the dinosaur finds included a new species of the long-necked sauropod *Mamenchisaurus.* This browser was the largest of Asian sauropods at an estimated weight of twenty-five tons and length of approximately one hundred feet, the longest-necked dinosaur yet known. With neck vertebrae about five feet long, thinner than a poster tube and rigidly linked, the neck was, according to Russell, "half as big as the whole animal," and double the relative size of most sauropod necks. What's more, it was preserved with a fidelity that astounded the Canadians. "The neck bones were like balloons, really light. But they were uncrushed," marveled Currie.

A large and new theropod was partially dislodged from the Sinkiang ground, this a highly unusual single-crested carnivore thought at first to be of the same species as a partial skeleton found by Chinese paleontologist Zhao Xijin in a 1984 Chinese expedition to the region (see Chapter 5). A complete specimen would have been twenty-five feet long. Zhao, a good-humored man of Dong's generation, was along to participate in the new excavations. As the new find was unearthed, skull and articulated foot bones first, Currie called the feet "the best preserved I've ever seen of any predator." As more of the crested theropod emerged, Dong, Zhao, and Currie became convinced this was something entirely new, a carnivore they have yet to name or formally describe.

The skull of a second large carnivore at the Sinkiang site was found just ninety minutes after Currie, Russell, and Dong set out on a thousand-mile jeep trek to scout sites in Inner Mongolia for the 1988 field season.

This arduous journey, a month long, was punctuated by many wrong turns. But as the group wended east, they found dinosaur bones scattered by the hundreds across the desert, including two *Velociraptor* skeletons, with a dramatic dune-swept canyonland marked by a sandstone column in the distance. The column had to be, Currie reckoned, the red tower of Ulan Tsonchi, approached by Andrews and noted by Sven Hedin, the Swede who had explored the Gobi a half century before.

The discovery of the leg bone of a large carnivore in Sinkiang, China, makes predatory-dinosaur expert Philip Currie a happy man. [Photo courtesy Ex Terra Foundation]

Dinosaur history was made here. The Flaming Cliffs of Mongolia where in 1924 the American Museum team led by Roy Chapman Andrews found the first nests of dinosaur eggs ever known. [Photo by author]

Bayan Manduhu

At Ulan Tsonchi, Currie and his collaborators didn't fare any better than the Swedes, Soviets, and Chinese had before them—they found nothing other than raging sandstorms. But twenty-five kilometers to the west, at the orange cliffs of Bayan Manduhu, they did much better. At that larger outcrop of the same formation that had yielded so many dinosaur fossils for Andrews further north at Mongolia's Flaming Cliffs, Currie and associates plumbed what proved to be the richest site yet in the Gobi for dinosaurs.

The canyons of Bayan Manduhu (ironically named "rich spring") provided more than twenty miles of continuous exposure of Late Cretaceous sediments along 150-foot-high escarpments. And within the sandstone were "bones everywhere," as Currie exulted.

As they sifted through the loose sandstone, the Dinosaur Project team was rewarded, almost instantly, with spectacular finds. In one day they came upon twenty-three skeletons of *Protoceratops.* Tyrrell Museum preparator Kevin Aulenback had barely started to dig when he found an embryonic skull of a *Protoceratops,* a beautiful specimen the size of a twenty-five-cent piece. Zheng Jiajian spotted an even tinier, perfectly preserved skull of a Late Cretaceous mammal, which was perched on a small rise. Noble was casually scooping away the sandstone, demonstrating prospecting techniques to some Chinese journalists, when he spotted several whitish fragments on the ground. He brought Currie over. "Brian thought he'd come upon a microvertebrate site," Currie recalls. Currie immediately recognized the distinctive leaf-shaped teeth of an ankylosaur, an armored dinosaur. All ankylosaur teeth are "incredibly small for the size of the animal, just a half-inch long in adults," says Currie. But these were much smaller, the size of a pinhead. "They were just three or four millimeters in size. I thought these might be baby ankylosaur teeth. I thought maybe we'd find a baby dinosaur. You try not to get your hopes up, but I was pretty excited."

As Currie dug further, he found a tail. The tail led to two heads. "Something very strange was going on." With a bit more probing with

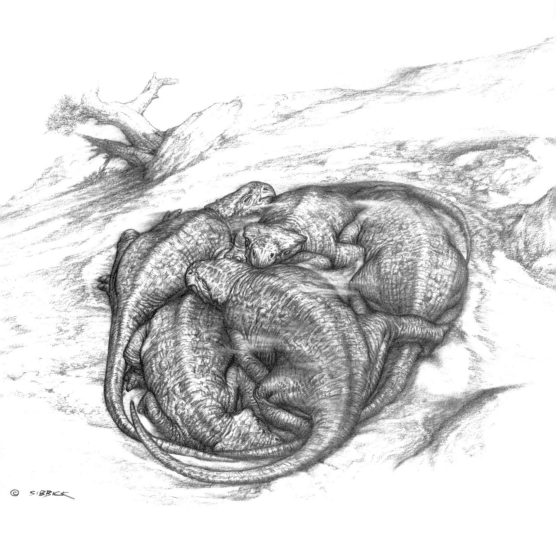

© SIBBICK

awl and fingertips Currie discovered five baby pinacosaurs, splendidly preserved. In adulthood these dinosaurs were near tank-sized with bony armor and tail-clubs. But these juveniles were the size of small sheep, just four feet long and too young to have developed clubs or armor plates.

Ankylosaurs were known previously from Asia and North America, but only as isolated individuals. Nothing was known of their social habits or development. But here, at Bayan Manduhu, baby pinacosaurs

---

clearly traveled together. They were probably a single family, hatched from one nest. And at the very spot where Noble had found them, they had died, pinned beneath a collapsing or shifting sand dune. Those at the bottom were nearly intact. Those closer to the surface of the dune had been preyed upon by a carrion-eating *Velociraptor,* which lost some of its loose teeth as it scavenged.

At Bayan Manduhu small mammal skulls, significant in themselves and reliable as relative indicators of the age of deposits (since the evolution of mammals is well documented), were found by the dozen. Partial remains of several other armored pinacosaurs were turned up, near oblong-shaped eggs like those Andrews had found at the Flaming Cliffs and thought belonged to *Protoceratops.* So Andrews may have been wrong about the identity of the egg-layers. "Perhaps," reckons Currie, "they belong to pinacosaurs."

The diggers uncovered the remains of hundreds of other dinosaurs, and crocodiles, lizards, and mammals as well. One tiny skull, only millimeters long, thought to be of a lizard, appeared upon preparation to be a fetal ankylosaur's, the tiniest dinosaur skull ever found.

Currie's preparator, Kevin Aulenback, discovered five different kinds of fossil eggshell from dinosaur eggs small and large, thick- and thin-shelled, spherical and oblong. He spent much of the following two years piecing together and studying the eggshells.

Paleontologist Paul Johnston found tunneling burrows some thirteen feet long. "At first we thought we were looking at sedimentary structures similar to stalagmites," said Currie. Small wonder they were confused. These "trace fossils" are the trails left by ancient insects— and Johnston believed those he'd found were the largest trace fossils ever known, left perhaps by beetles feeding on dinosaur dung.

That first week of serious exploration of the canyonland past the Red Tower had yielded more than 125 distinct and significant finds, the greatest number of animals from dinosaur time ever found in one Chinese dinosaur locality.

Bayan Manduhu was of special import not only for what new information it provided about Mongolian dinosaurs and their behavior, as well as their resemblance to dinosaurs of North America (particularly small predators and the ankylosaurs), but also for the information the site provided about dinosaurs' remarkable adaptability. For Bayan Manduhu was very nearly a desert in dinosaur days, a habitat entirely unlike the subtropical deltas occupied by many North American dinosaurs of the time.

Sentiments From Sediments

Reading the ancient environment is a sedimentologist's work. Sedimentologists and stratigraphers only occasionally accompany dinosaur paleontologists on fossil hunts. Most paleontologists rely instead on previous geological studies or on their own informed sense of the former lay of the land.

But reading the landscape that hosted the greatest collection of Asian dinosaurs yet was a task that fell to a fiftyish Polish veteran of the 1971 Polish-Mongolian dinosaur dig, Tom Jerzykiewicz.

He'd waited more than a decade for the chance to substantiate a particular analysis he'd made of the most extreme and dramatic of Mongolian terrains, that of the Flaming Cliffs. Now, unexpectedly, he'd come upon a canyonland that replicated the same conditions in the same formation, on an even grander scale.

Like the other dinosaur-bearing rock formations of Mongolia, ranging in age from 80 million to 65 million years old, the Djadokhta Formation of Bayan Manduhu and the Flaming Cliffs is built upon a geological platform called a tectonic graben. *Tectonic* refers to the movements of the earth's crust, sheets often broken near their surfaces by faulting—splits in stressed rock. A long, narrow block of land that sits, sunken, between two parallel blocks of the upthrust fault rock is a graben. The upper Rhine Valley is a graben, as is the East African Rift Valley.

The center of one such graben, where Bayan Manduhu lies, rose in the Late Cretaceous with a slow accumulation of sedimentary rock. The accumulation of sediment was interrupted many times by long and arid intervals when rivers no longer washed down debris into the lowlands. Instead, desert winds were rearranging and removing the existing landscape, and creating the windblown dunes that sometimes collapsed on baby dinosaur such as the armored dinosaurs of Bayan Manduhu.

That environmental portrait painted by Jerzykiewicz ran counter to the sense of the Polish and Soviet geologists who had studied the Djadokhta before him and called it a lake-dominated environment (hence the notion of quicksand trapping the so-called "Fighting Dinosaurs" found by the Poles). Theirs was a logical conclusion, as even into the 1970s there was no other means known for the accumulation of limestone in soil profiles than from lake-bottom deposits, and limestone is indeed abundant at Bayan Manduhu and the Flaming Cliffs.

But Jerzykiewicz knew of a more recently detected means for limestone to become locked in the ground. During arid intervals primitive soils begin to develop, and a form of limestone, calcium carbonate, accumulates in nodules within the soil by erosion and the weathering processes of soil formation.

The conditions needed to trigger this process are now known with impressive exactitude. Roots and stems can concentrate minerals. Under the right conditions—warm, semi-arid conditions of sporadic rainfall-nodules of calcium carbonate, probably attached at first to plant roots, can build up in the ground. Too much rain and calcium carbonate will leach out of the soil layer. Too little and it will not build up. But in semidesert environments of about a foot and a half of rain a year, a nodule zone can build. Follow this deposition with long periods of drought and a soil-like layer of limestone can be formed. This process has recently been studied in detail in modern soil profiles in New Mexico and southern Texas.

Jerzykiewicz had sensed that the Djadokhta Formation limestone might have built up in such a manner when he saw the Flaming Cliffs in 1971, but as the knowledge of such processes was young, he held his view in check. At Bayan Manduhu as at Flaming Cliffs, he saw signs of a highly arid ancient environment. At least eight narrow bands of whitish yellow limestone, sandwiched between the orange sandstone dune deposits, suggested to his eye that this area, and its dinosaurs, had withstood drought conditions over periods of more than a hundred thousand years. Such extended periods of aridity would be necessary, he reckoned, to deposit and act upon these mature soillike layers of limestone. Unlike the limestone deposits left at the base of lakes, these limestone layers did not contain lake-bed fossils.

And the sandstone as much as the limestone at Bayan Manduhu revealed an ancient semidesert environment to the sedimentologist's eye. The sandstone was formed of powderlike grains. Had lakes or rivers shared this terrain, their waters would have separated its finer particles, the clay and silt, from the larger particles of sand. But here "all fractions were well mixed," Jerzykiewicz noted, thoroughly jumbled by eons of windy drought. On the lee side of the Mesozoic dunes, a cross-stitching pattern was plainly visible. The sand had been laid down in this fashion by shifting winds, and by ephemeral streams emptying into small, seasonal ponds, not by constantly flowing or meandering streams.

The sedimentologist's contribution to dinosaur paleontology was clear in Mongolia—Jerzykiewicz provided the evidence that some smaller dinosaurs dwelt in harsh, arid conditions. And he considered the information equally useful to sedimentology. "Now we can say that the presence of certain dinosaurs is indicative of semidesert environments."

A comparatively meager variety of dinosaurs inhabited this difficult environment. Though Bayan Manduhu produced an unprecedented number of fossils for a Chinese site, they belonged only to small to medium-sized vertebrates. The largest, *Pinacosaurus* and *Protoceratops,* grew no bigger than eleven feet long. Says Currie, "They're good-sized animals, skulls up to a meter across, but not the biggest of dinosaurs. Perhaps it was the shortage of vegetation that would not support more large animals."

And how had the Americans, Swedes, Russians, and Chinese, all having come so close at Ulan Tsonchi, missed Bayan Manduhu, one of the best of all Mongolian dinosaur dig sites? Says Currie, "The dune sand may have been too soft for their vehicles to continue farther west. And the beds at Ulan Tsonchi did not give them much hope, anyway." Without thorough ground or aerial surveys, none of them could have known what lay nearby.

Iren Dabasu

New methodology as much as new technology account for the insights paleontologists now obtain even in locations where their dinosaur-hunting predecessors have already been. Such was the case at Iren Dabasu, where Andrews had made his first dinosaur finds.

The 250-mile trip from Bayan Manduhu to Iren Dabasu (now Erenhot) took the Canadian-Chinese team ten days. The caravan of jeeps and trucks was often mired and lost in the deeply rutted and unmarked caravan trails. At one point they were waylaid by local authorities who asked exorbitant fees for road improvement taxes. And the team was delayed by courtesy calls on local officials that devolved into all-night feasts.

At Erenhot more fossil pieces were added to the dinosaur migration jigsaw, a bone bed of hadrosaurs akin to those of North America, a small *Troödon*-like hunter, and many eggs. On the final morning of prospecting, true to the "last day phenomenon" of fossil quarrying, Currie and his crew came upon their biggest find at Erenhot. It was a partial skeleton of an *Alectrosaurus,* an uncommon midsize predator from Central Asia, with an intact wishbone. "No one's reported a furcula for an *Alectrosaurus* before," Currie enthused (he discovered when he journeyed to Mongolia the next summer that the Mongolians had already made such a find). No sooner had he and Aulenback whisked the dust off the elegantly curved white bone than they had to cover it over again. They brushed a mound of dirt over the bone and set an orange flag on a wire beside it, Aulenback noting the field coordinates in his notebook.

Many small fossil fragments were concentrated by erosion of the mounds left by the bulldozing Russian scientists nearly thirty years before. Among the bits of bone were several isolated predatory dinosaur teeth, the teeth of uncommon tyrannosaurid dinosaurs. Nonetheless, Currie found them "real easy" to recognize. "They all have serrations, front and back of the tooth, of a peculiar shape. If you pick one serration, you'll see it's chisel-shaped. At the base of the chisel are grooves that go across the teeth, so they won't bind in flesh the animal is chewing."

The discovery of various parts of hunters and browsers at Iren Dabasu, dinosaurs known in North America from the end of the Cretaceous, suggests that these Mongolian dinosaurs were near contemporaries of those Currie knew from Alberta. These finds, like those of Bayan Manduhu, gave fresh evidence that there was a large-scale dinosaur movement between the continents after all.

Currie found better evidence of Asian-American dinosaur migrations in still-unpublished Albertan finds. Those fossil discoveries reveal that at least two of the three kinds of pachycephalosaurs that Kielan-Jaworowska's team found in Mongolia and thought unique to Asia are found in North America as well. Kielan-Jaworowska had suggested mammals moved across the continents but dinosaurs did not. But her colleague, Polish paleontologist Halszka Osmolska, is now convinced of the opposing view, that "there was an exchange of faunas between Asia and North America during the Late Cretaceous, at least of dinosaur faunas—the mammals, for some reason, did not migrate." Overall, the case for an Asian-American dinosaur migration is now compelling.

But as is often the case, all the dinosaur evidence did not neatly fit the theory. Some of the discoveries produced false leads on the trail to drawing correspondences between Asian and North American dinosaurs. Among Currie's finds were six distinctive foot-bone fragments he identified as being from *Elmisaurus rarus* ("rare foot lizard"), a fossil animal so rare, as its name implies, that Currie's Iren Dabasu haul equaled the previously known worldwide take. This dinosaur's feet have three long toes and a short "big" toe with several of the foot bones fused. Half of the specimens found previously were from Mongolia, half from Alberta, which added strength to Currie's theory that "all the small theropods are basically the same in Late Cretaceous North America and Asia."

However, by some impressive detective work, Currie proved himself wrong about his new *Elmisaurus* find. He compared the bones to known *Elmisaurus* feet in Canada and Asia and to two specimens wrongly labeled as such in the American Museum collections from the Andrews expedition. He visited the Soviet Union and examined *Avimimus,* another small birdlike predatory dinosaur known only from Asia, and discussed his find with Soviet paleontologist Sergei Kurzanov, the describer of *Avimimus.* Only then did Currie conclude that what he'd found were relatives of *Avimimus,* not *Elmisaurus.* "So they say nothing about migration, but plenty about dating," says Currie of his error.

Avimimus-like dinosaurs are known only from the latest Cretaceous, so the previous hazy dating of the Iren Dabasu fossil beds was placed in further doubt—these dinosaurs were indeed later than paleontologists had long thought, and nearer in time to those known well from North America, 75 million years ago.

Not all dinosaurs from Late Cretaceous Asia have known cousins in North America and vice versa. But the biggest unresolved question in attempting to link North American and Asian dinosaurs is the absence in Asia of horned dinosaurs, so plentiful in the American West. Perhaps the horned dinosaurs were not adapted to grazing in semiarid country, as protoceratopsians were. The *Protoceratops* dinosaurs of Mongolia do have relatives in Montana, recently found in what were drier and more upland habitats than those common for many North American dinosaurs, and closer to the harsher conditions of the ancient Gobi.

So, in time most of the differences between Asian and American dinosaurs in the Late Cretaceous may be explained by habitat distinctions and preferences. Asian plant-eating dinosaurs may be less diverse because they inhabited drier, poorer terrains. If they appear highly distinct from North American forms, it may be because paleontologists have yet to investigate many of the more arid and Asianlike American habitats.

With these qualifications in mind, some tentative conclusions can be drawn about dinosaur migration between Asia and North America. Carnivorous dinosaurs were more alike across continents and more diverse than herbivores in Asia. Why? Currie offers two reasons. Hunters are more mobile than their prey, quicker to exploit new areas. And carnivores would, Currie expects, tolerate "a wider range of environments. To them it doesn't matter where they are, what they're with, as long as there are dinosaurs to eat."

Asia may yet prove to be the motherland of many dinosaurs, as Osborn and Andrews contended. From as much as can be discerned from shaky dating and scattered sampling of fossil-bearing sediments, there are no known North American counterparts to the parrot-beaked psittacosaurs of China and Mongolia, ancestors of *Protoceratops* and *Triceratops* alike.

Divining the origin and movements of dinosaurs across the Asian and North American continents is a large order indeed. Now that the Chinese-Canadian dinosaur hunt has ended, perhaps answers to the migration question will come from the joint explorations of the American Museum and Mongolian scientists in the Gobi, begun in 1990, or from a cooperative effort of Inner Mongolian dinosaur scientists and Paul Sereno, begun in 1991. In the same Gobi valley where the Poles had enjoyed success, American Museum scientists began digging in 1990. In just a few days of prospecting, they found a spectacularly well-preserved young *Protoceratops* skeleton, splayed out like a tiger-skin rug on a hillside. Over another rise they found a dinosaur nest with half a dozen intact eggs. And on a third hillside, a fifteen-foot ankylosaur skeleton spread out on its back. Nearby, American Museum paleontologist Malcolm McKenna came upon something entirely new, sparkling white on the cliff-face—a seven-foot-long Komodo-dragon-like lizard from late in dinosaur time.

In the summer of 1991, the American Museum crew returned for nearly three months across 2,500 miles of the Gobi. They discovered many more fossils, including exquisite skeletons of several new species of small predatory dinosaurs, relatives of *Dromeosaurus, Velociraptor,* and *Oviraptor.* And at the Flaming Cliffs they discovered another *Protoceratops* nest, just like the historic find of the Andrews expedition seventy-six years before. They will no doubt find more in summers to come.

One of the smaller treasures of the Mongolian State Museum is this complete skeleton of a baby duckbilled dinosaur, perhaps the most complete of its kind in the world. [Photo by author]

Many questions remain unanswered and many discoveries remain to be made before the age, origins, and environment of the last Asian dinosaur is understood. Says Currie, "We may not have got it nailed down yet, but we're two or three steps closer to a solution."

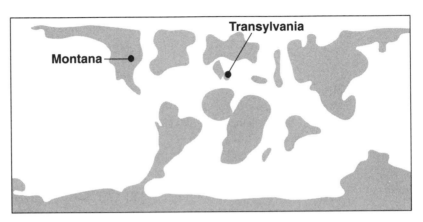

The breakup of the supercontinents, 80–68m. yrs. ago

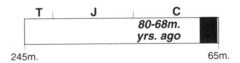

T	J	C

**80-68m.
yrs. ago**

245m.   65m.

AGE OF THE DINOSAURS

Triassic Jurassic Cretaceous

First animals   First land animals   First human

| | | | T | J | C | | |

590m.   350m.   245m. 208m.   145m.   65m.   Today

CHAPTER 10

Eggs, Babies, and New Dinosaurs

From a distant hilltop, the western Montana vista is one from centuries ago—a half dozen brightly colored tepees, nestled along the banks of a rushing river, the only habitations in sight on badlands and rolling grassy hills.

The hillsides, and the cliffs and canyons of the fast-eroding badlands, haven't been here for much longer than the Indians who now own the land. The last ice age, forty thousand years ago more or less, scoured this land and exposed the arctic bridge that brought the Indians.

Recently, Jack Horner discovered the 75-million-year-old contents of these hills—the world's richest known supply of duckbill dinosaur fossils. His crew of eight digs here all summer, scraping and brushing the black bones of dozens of scrambled duckbill skeletons that lie exposed, painting them with clear preservative, and plastering them for shipment to the Museum of the Rockies in Bozeman.

Horner doesn't help. Instead, he is off walking. What Jack Horner sees on his long walks across the badlands of western Montana isn't clear to others. Since he goes alone and doesn't talk much when he comes back, his methods of observation remain a mystery.

Unquestionably, he sees dinosaur fossils, in an abundance and variety few have ever seen before. He finds them because he sees this land, with singular clarity, for what it is, a readable geological record of Cretaceous habitats near a huge and shifting continental sea. Mapped in his mind are scenes like this:

From a forested upland where the duckbill dinosaurs spent spring and summer hatching and tending their fast-growing babies, a vast herd proceeds by long, two-legged strides, south along the highlands. Heads bob as they slowly march. Elaborately ornamented males stride in front, issuing rumbling toots from their nasal tubes to keep stragglers in line. Other adults guard the flanks of the herd and the half-size juveniles at its center. All are alert to the sight and smell of packs of small hunters and the huge and dreaded tyrannosaurs.

Browsing as they head to the subtropical shore, the duckbills pass other residents of this productive land: bounding herds of lithe, little hypsilophodontids, and clusters of squat, young armored dinosaurs. And the migrants skirt large herds of horned dinosaurs, whose huge, bony frills point upward as they noisily munch, faces in the brush.

For his discoveries and insights into just one aspect of this complex social scene—the birth and nurturance of duckbill dinosaurs—Horner has become famous as the dinosaur "egg man." But his objectives go even further than collecting evidence of the life, growth, and behavior of this dinosaur community on a scale never before approached. He and colleague Dave Weishampel of Johns Hopkins University are in the midst of an evolutionary study of unrivaled scope and sophistication in dinosaur studies—an attempt to demonstrate the effects of changing habitat over millions of years on the rate and manner in which new species were formed.

Toward that end Horner can be found working as he does every summer, in remote spots much like this one, all of them not a hundred miles from the town where he was born, a sleepy railway stop called Shelby, in the midst of the immense Montana flatland.

Herds of thousands of maiasaurs traveled the western plains in search of forage 75 million years ago. [Donna Braginetz]

Since he was eight and found—and numbered—his first dinosaur bone on his father's ranch, Jack Horner has known exactly what he wanted to do with his life: "study duckbills." Not simply paleontology nor specifically dinosaurs. "Nope, just duckbills."

Horner didn't distinguish himself in school, but then none of his teachers recognized the twin roots of his learning problems: dyslexia and boredom. When he returned to the U.S. from Vietnam with GI benefits, Horner decided to pursue what interested him—duckbills. He took exclusively paleontological courses at universities from Montana to Princeton, and at the latter he landed a job as a fossil preparator.

Paleontologist John R. "Jack" Horner crosses many obstacles in his search for duckbilled dinosaurs in Montana. [Photo by author]

In 1982, Montana State University in Bozeman offered Horner the opportunity to return to Montana, as an assistant curator of the Museum of the Rockies, at a very modest salary. He eagerly took the post and remains a curator at the museum. He teaches a course or two through the long winter, organizes the museum's collections, sections bone, writes papers, prepares fossils, and waits for his fossil preparators to dust off the summer's finds so he can study them in detail.

And it is an irony of paleontology's unpredictable fortunes that the fossils Horner prepared at Princeton are no longer at that wealthy and prestigious university (they were given to Yale University). Montana State, where Horner now works, boasts one of the newest and finest dinosaur galleries in the nation, composed chiefly of models of Horner's finds, and the first laboratory devoted to dinosaur-bone histology in the United States (see Chapter 6).

The changes in Horner's own life have been equally dramatic in the past decade, though they haven't affected his taciturn, defiantly unpretentious personality. Not until 1986 did he and his wife move out of his small, graduate-student hovel and into their own home. Their move was made possible by an unsolicited, unrestricted five-year $200,000 "genius" grant to Horner from the MacArthur Foundation. "When I called Jack to tell him he'd won," recalls former museum director Mick Hager, "his response was, 'Gee, I've got to walk around and think about this.' That's all he said."

The award, in turn, was based on the solid and celebrated research Horner has done since he returned to Montana. In the public's mind, Horner is a fossil finder. "I'm the guy who digs them up, Bob Bakker's the guy who figures them out," says Horner with considerable irritation. Horner did find dinosaur eggs, the first eggs and nests known from North America. He also found and continues to find an astounding array of fresh information about how duckbills lived and how they and other dinosaurs evolved. But he's found this all out because, as several paleontologists have said, Peter Dodson among them, "Jack is a very careful scientist. The first time he told me about maternal care of baby dinosaurs in nests, I thought he'd gone off the deep end, but he can show you the evidence."

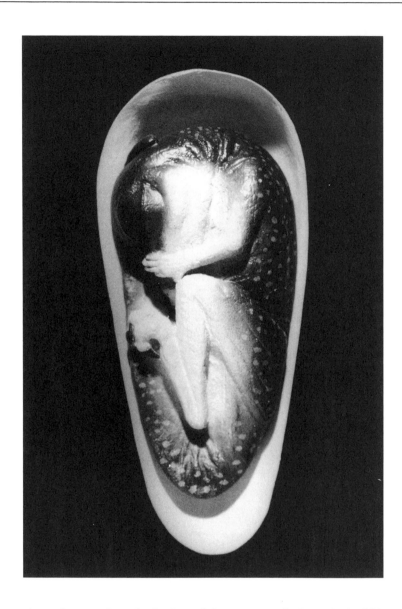

An embryonic hypsilophodontid dinosaur might have looked like this sculpture, fashioned by Montana sculptor Matt Smith from dinosaur bones found within unopened eggs by Jack Horner. [Photo courtesy Museum of the Rockies]

In Horner's mind his fossil-finding talent is simply a matter of knowing what to look for. In 1977, a year before he found not just eggs but a nest filled with fifteen dinosaur eggs and the bones of many babies, Horner told friends he would be finding eggs and babies the next season, based on what he'd learned of the western-Montana landscape.

What Horner had learned, by then, was the lay of an inland sea that ran from the Arctic to the Gulf of Mexico, through what was, in Cretaceous times, land as warm as Florida is now. He had reasoned, as fossil hunter Charles M. Sternberg had suggested many years before, that uplands were a good place to look for juvenile dinosaurs. And in 1975, Dodson had reidentified purported adult dinosaur specimens from the area as juveniles.

Horner reasoned further that duckbills fed in the lowlands but came away from the egg-damaging acid coastal soils to drier uplands. "Like birds, they made huge nesting colonies," says Horner.

So Horner looked to the Cretaceous highlands, a hundred miles or more from the sea, and with some serendipity came upon the site later famous as "Egg Mountain," a windswept scattering of hills and gullies near the sleepy town of Choteau, sixty miles southwest of Shelby and within sight of the Rockies. Before he knew of Egg Mountain, Horner and his best friend and crew chief, Bob Makela, were shown a coffee can full of pencil-thin little fossils in a Bynum, Montana, rock shop by its owner, Marion Brandvold.

Horner recognized the little fossils at once as baby duckbill bones. On low gray mounds in the midst of the tabletop-flat Peebles' ranch, where the Brandvolds had prospected, Horner and Makela found the first nest of baby dinosaurs ever discovered. The following year Horner found more of these duckbill nests and eggs.

With the duckbill eggs, Horner also found more to astound anyone but him—bones of nestlings, juveniles, and thirty-foot-long adults, fossilized dung, mounded plant material of nests, and more well-chewed matter that Horner believes may have been regurgitated food, which mothers fed their young. Horner had found not only babies, but the evidence to piece together a complete portrait of these duckbills' development.

The traces of many blood vessels in the bone indicated these young grew fast. But Horner saw several indications that these babies, like many modern birds, required extended parental care and feeding. He found broken eggshell in many nests, indicating the young had remained in the nest long enough to trample their eggshells. Unfinished bone surfaces on the tiniest young showed the hatchlings were not equipped to travel far or to fend for themselves soon after birth. And the fossilized regurgitant and dung Horner saw in and around the nests suggested the duckbills' parents were on hand to feed their young. With that care-giving in mind, Horner named the previously unknown duckbill *Maiasaura,* the "good-mother lizard," the first dinosaur with a feminine title.

At Egg Mountain in 1979, Horner's crew began unearthing many eggs and nests of ten-foot-long or smaller vegetarians, hypsilophodontid dinosaurs. Four years later, at the Egg Island site close by, his crew found a nest of nineteen hypsilophodontid eggs with tiny bones preserved within. And again nearby, he and his crew found seven duckbill eggs, these also with intact embryos.

As luck is the residue of design, bones seem to appear everywhere Horner goes. When he and his crew pitched their tepees and tents in a shallow depression near Egg Mountain, they found the ground particularly unyielding. Digging away at the stake-holes, they uncovered more dinosaur bones, thousands of them. The fortuitous find of the jumble of duckbill bones gave the subsequent excavation site its name, "Camp"-osaur Quarry.

Horner gave up Egg Mountain in 1984, though he returned briefly to excavate *Troödon* eggs and bones. (The Nature Conservancy purchased the site, and the Museum of the Rockies runs a program of volunteer digs there every summer.) He left to scout more northerly sites he'd long had his eye on. These locales had been explored for dinosaurs in a far more superficial manner long ago. At one prospective site, along the Two Medicine River, Horner met an elderly local who remembered seeing a paleontologist on the land, Barnum Brown, the American Museum's chief dinosaur digger nearly a century ago.

Horner also located a spot on Blackfoot Indian land, near the Canadian border, that Charles Gilmore of the Smithsonian had briefly quarried in 1928. Horner called the place Landslide Butte. At the base of the grassy butte, down a steep, rutted dirt path, Horner pitched his camp. Across a quiet stream the folds and stripes of the beige and brown badlands were in plain sight. On and in those hills are dinosaur bones, by the millions.

Among the treasures scattered throughout the badland mounds near Landslide Butte are sharp, hooked, and serrated inch-long teeth of small predators, and small fragments of dinosaur bone, colored slate blue, dark red, and gray by minerals that have worked their way into the bone's interstices. Here, too, dinosaurs nested. Walk anywhere on these hills and gulleys and one will see bits of black eggshell. Digging gently into clifftops with spades and awls, Horner's crews have come upon hadrosaur leg bones nearly as large as a man. They have found complete skulls of unknown ceratopsians with single drooping horns (earlier relatives of the well-known three-pronged *Triceratops*). These skulls were so massive that once their jackets of burlap and plaster were applied, they had to be helicoptered off the site.

But those finds only begin to describe the fossil wonders of Landslide Butte. As Horner wrote, "The Willow Creek anticline [where the *Maiasaura* and hypsilophodont eggs and a hadrosaur bone bed were found] had been a surprise. Landslide Butte was a shock."

Here was a site so rich in fossils that Horner has estimated 53 million bone pieces can be found in the space of a few square miles. In this spot duckbills nested and died by the tens of thousands, killed suddenly by a local catastrophe, likely a drought. Horner cites a contemporary analogy: "Ever see how carcasses pile up around a dried watering hole in the Serengeti?"

In just a few square miles of these barren hills, Horner found two huge horned-dinosaur bone beds and three bone beds full of duckbills. He found the first horned dinosaur nesting ground. And, he writes, "in the most astonishing discovery, we found a nesting ground of hadrosaurs that is a mile wide, three miles long, and three horizons deep. That is to say, stacked on top of one another are three nesting grounds of this one dinosaur, each three miles square." Where Horner had to crawl on hands and knees to find indications of nests at the Willow Creek anticline, at Landslide Butte "there are spots where,

Museum of the Rockies crew chief Pat Leiggi plasters the leg bone of large duckbilled dinosaur. [Photo by author]

without even digging, you can literally shovel up the baby bones."

To his surprise, Horner's ongoing discoveries of eggs, nests, and babies, only a fraction of which he's yet published, aroused enormous public interest. When the governor of Montana took a cast of a *Maiasaura* egg on his state visit to Japan, fifteen thousand people came to see it on the first day of its exhibition.

Horner's discoveries have sparked unprecedented discoveries of dinosaur eggs and babies by his colleagues around the world.

The Egg Men

Dinosaur eggs were reported first in the early 1920s from southern France and from Roy Chapman Andrews's adventures in Mongolia. Clutches of whole eggs have been located from deposits in many Chinese localities for decades, and the Chinese have even named at least a half dozen different dinosaur species by their differing eggshell structures.

The first North American dinosaur eggshell was reported from fragments found in the Hell Creek Formation of eastern Montana in 1930. Earlier, in 1928, Charles Gilmore had found eggshells at Landslide Butte, but he hadn't thought to publish on them. In 1966, Jim Jensen wrote up Late Cretaceous dinosaur eggshell he'd found in Utah.

But suddenly, in the past decade, eggs have been turned up around the world in relative abundance. In Mongolia, the finds have been especially rich. The Chinese-Canadian expedition found five kinds of eggs in 1988, including elliptical ones identical to those identified by Andrews's expedition members. (Mongolian paleontologist Artangerel Perle recalls finding thirty-five eggs on his first field trip in 1969, forty-four at another site, and a hundred more elsewhere, but word of the discoveries did not reach the West until the 1980s.) Nests and intact dinosaur eggs, some with embryonic bones still inside, have been found in India and Argentina, South Africa and Transylvania.

In Montana, Horner's own excavations of whole or partial eggs now numbers "seven hundred or eight hundred" by his own estimate. After nine years of Horner's gentle prodding, his friend Phil Currie looked for eggs on the Canadian side of the border in 1987 and found them in abundance, more than a dozen nests with thirty eggs in each. For Currie, as for Horner, the right area has proved to be the fossil-preserving sediments from drier, less acidic upland habitats.

The dinosaur eggs excavated from these horizons come in many sizes, though all are smaller than a foot long or wide. But they appear to sort readily into two shapes—round and elliptical.

The round, the size of large grapefruits, seem to belong to "lizard-hipped" dinosaurs (the huge sauropods in particular). The small elliptical ones, from four inches long to twice that size, appear to be characteristic of "bird-hipped" dinosaurs (duckbills, horned dinosaurs, and hypsilophodontids, though the one certain predator's egg, which Horner found of the "lizard-hipped" *Troödon,* curiously fits this category, and Canadian duckbill eggs appear to be more globular in shape).

From Horner's finds, and studies of them by eggshell expert Karl Hirsch, it appears at least some dinosaur eggs can be distinguished not only by size and shape but by patterns on the outside of the shells. The elliptical *Maiasaura* eggs had crinkly surfaces. The four-inch-long and narrower *Troödon* eggs were pebbly in texture. And the five-inch-long *Orodromeus* eggs, pointed on one end, blunt on the other, were smooth, and thinly striped in a faint lengthwise pattern.

Further clues to the identity of the egg-layers come from the arrangement of eggs in the nest. "Bird-hipped" dinosaur eggs appear to have been laid in spiral clutches. Nests such as those in Mongolia and in Montana show a circle of a dozen or more closely packed eggs, angled into the ground. Several broken ones have revealed duckbill and hypsilophodontid embryos in their shells.

The identity of sauropod eggs and nests was less certain, due to a paucity of evidence (one associated sauropod and its bowling-ball-sized egg come from southern France, reported in 1957), until a recent spate of finds. Eggs of the same sort as the French discovery have recently been found in northern Spain. The animal thought to have produced those eggs is a little-known, small sauropod from poor outcrops. This purported sauropod's best-preserved crushed egg, with an estimated volume of half a gallon, is the biggest dinosaur egg known, not much larger than an ostrich egg and only a fraction the size of the egg of *Aepyornis,* the elephant bird, a relatively recently extinct giant bird.

The spherical French eggs were arranged in two straight lines, like the children's book character Madeline and her little friends. This pattern now seems characteristic of sauropod egg-laying since the design is seen in multiple nests of intact sauropod eggs lately found by researchers in Patagonia.

Dinosaur eggs are now known from almost the entire breadth of dinosaur days. The earliest dinosaur eggs, likely from a prosauropod, may be those found in Africa, from the Red Beds of Lesotho and South Africa. In these deposits, now thought to be nearly 200 million years old, a clutch of long oval eggs was discovered in 1977 by the legendary fossil finder James Kitching. Or the oldest dinosaur eggs may be from Argentina, where Jose Bonaparte reported finding two little eggs with the little *Mussaurus* dinosaur from the Late Triassic, whose fossil bodies curled within his palms. Eggs from some of the last of the dinosaurs were found by a Chinese researcher, who observed in 1979 that these eggshells were thinner than those from older localities, suggesting to him that thinning eggshell was related to dinosaur extinction.

Detailed study of dinosaur eggs may provide clues to dinosaurs' identity, their evolution, and their body chemistry. Dinosaur eggs are much like bird eggs in their microscopic structure, enough to make the two difficult to distinguish, as opposed to the markedly different structures in the eggs of lizards, turtles, geckos, and crocodiles. Dinosaur eggs appear to be differentiated by at least four distinct variants of canals and structures, all prospective clues to sorting out not only which dinosaur made which egg, but which dinosaurs are the closest in their evolutionary histories.

Digging Up Baby

While the study of dinosaur eggs and eggshells is booming, what is truly unprecedented in dinosaur paleontology is the discovery of embryos within dinosaur eggs. (Andrews reported some in his *Protocera-*

tops from Mongolia in the 1920s, but as recent X rays have shown, he was mistaken.)

Jack Horner has found more than two dozen embryos, seven from *Maiasaura,* more than a dozen from a hypsilophodontid, one from a *Troödon.* Big-eyed and small-chinned, they hatch looking "cute," a subjective but not unscientific appraisal, and one that, Horner suggested at the 1988 SVP meetings, contributed to the care they received from their parents. They are also so complete and completely known by Horner that they have served to distinguish the different ways coeval dinosaurs grow up, even to identify new forms of dinosaurs.

In the spring of 1988, Horner and Weishampel published the results of their analysis of the embryonic contents of a five-inch-long egg, found lying on its side among a clutch of eighteen other upright unhatched eggs at Egg Mountain in 1984. While doing X-ray examinations of all the eggs later that year, Horner found that the sideways egg contained an intact embryo, curled in fetal position.

Using minute picks and a hundred-power microscope, Horner spent much of the next four winters painstakingly separating one side of the fetal skeleton from the rock that filled in the egg's interior.

Horner and Weishampel concluded that this embryo represents one of the swift little hypsilophodontids, but of a new genus and species they named *Orodromeus* ("mountain runner") *makelai* (for Bob Makela). Even in the embryo, the "bones showed finished articulation surfaces," according to Weishampel, an indication that members of the species were nearly independent when they hatched. "They could have hit the ground running," says Weishampel. "The young may have hatched out and scampered together outside the nest" as penguins and ostriches do, Horner suggested. "That's probably why we find intact eggs—the parents and the young aren't in the nest stomping on them."

Among embryonic and even hatchling maiasaurs found by Horner, bone surfaces are pitted and far from smoothly finished, indicating those animals required extensive care and feeding after birth.

Since both the hypsilophodontid and the maiasaur embryos and baby bones show the woven pattern of primary bone associated with endothermic animals, one might well conclude, as Horner has, that these dinosaurs were fast-growing, perforce warm-blooded. The hadrosaurs would have needed to grow fast to be big enough to migrate with the herd in a few months' time after hatching. The hypsilophodontids needed to be fast and fast-growing to escape the quick and cunning predators, such as *Troödon,* that pursued them. These dinosaurs were hounded, it would seem, almost from birth, given the presence of *Troödon* eggs in hypsilophodont nests. The young and helpless maiasaurs might have been threatened by *Troödon* as well, but they enjoyed the protection of attentive parents.

The indicated rate of dinosaur growth affects not only our estimates of their metabolism but of their sexual maturation and of how long dinosaurs may have lived. Bigger animals tend to live longer than small ones. Some dinosaurs may have lived well over a hundred years, for based on reptilian growth rates, the giant sauropods would not have reached maturity until their second century.

Figuring in a mammalian or avian growth rate, as suggested by Horner's baby dinosaur finds, young of even the biggest dinosaurs would have been ready to breed in a decade, or even earlier if they had the reptilian trait of breeding at sizes well below maximum adult size.

Embryonic teeth provide other behavior insights. The slice-and-dice teeth on the in-the-egg hypsilophodont indicated to Weishampel that the animal ate "either pulpy leaves or fruit." The teeth lacked the coarse edges found on more mature hypsilophodont tooth fossils discovered at the site. "It may well have eaten a completely different diet while young," said Horner. "Many animals do that."

From the Milk River, Alberta, site of his discovery of eggs, embryos along with hatchlings, juveniles, and adults, Phil Currie and Tyrrell crews have extracted a rich load of information on dinosaur development. From two eight-inch-long eggs the researchers removed bones and used them to make a composite skeleton. The delicate black bones are aligned to form a one-and-one-half-foot-long embryo. The baby's domed head is immediately recognizable, at least to a dinosaur paleontologist, as the skull of a crested duckbill. Adults of the animals grew to thirty feet, but these embryos fitted into eight-inch-long eggs. Some of the fetal tooth surfaces were worn smooth, contradicting previous suggestions that tooth wear was one way to discriminate between hatchlings and embryos. Dinosaur embryos apparently gnashed their teeth. In one nest the embryos were 30 percent larger than those in the other nest, suggesting the bigger duckbills were nearer to hatching. As they were found in two different layers of sediment, Currie believes the site was used repeatedly, perhaps over thousands of years, as a nesting site.

Still more clues about the development of dinosaurs will come from the newly sophisticated CAT-scan images of dinosaur fossils. In the hands of technically expert researchers, such as Andrew Leitch, the CAT scanner and its 3-D imager companion are particularly well suited to examining dinosaur embryos. Without opening the delicate fossil eggshells, the contents of the sealed eggs can be examined. Bones can be distinguished from sediment that may have leaked into the eggs. And if the young weren't old enough at death for their bones to have hardened, cartilage can be discriminated. If there is an embryo inside an egg, any of its features can be examined in three dimensions, rotated and expanded at will.

The Last Dinosaur Hurrah

Although dinosaur eggs and embryos are a great interest of Horner's they are only a part of his scientific mission: furthering our understanding of what is now the best-known community of dinosaurs in the world—the animals of the North American West some 75 million years ago. "We're doing paleoecology here," Horner says. "We want to understand the whole environment—the plants, the animals, and how they got here."

Judging from our records of dinosaur diversity, this was the best of all times for dinosaurs. Our current understanding of this dinosaurian golden age is the result not only of Horner's studies in western Montana, Currie's in nearby Alberta, and Dodson's in both places, but of nearly a century of dinosaur-digging in these lands. As a result, these dinosaurs, and duckbills and horned dinosaurs in particular, are the best understood of any dinosaur groups.

Paleontologists have found the complete skeletons of adult duck-billed dinosaurs, down to skin and ossified tendons. Where skulls are hard to come by for most other dinosaurs, complete skulls have been found of nearly all the duckbills. What's more, thanks to Horner and others since, we have intact duckbill eggs with embryos, hatchlings and juveniles, and their nests.

Hadrosaurs—the duckbilled dinosaurs—grew large, some over thirty feet long and 6,500 pounds. They flourished in the Late Cretaceous not just in North America but in eastern Asia, Africa, Europe, and to a far lesser extent, South America.

Among the features that set the duckbills apart from other dinosaurs are their broad beaks, long low skulls, and teeth organized in a complex array known as a "dental battery." Their faces have been reorganized from those of their iguanodont ancestors to make room for a horsey snout full of teeth and among many duckbills, a solid or hollow crest atop their heads.

The duckbills are themselves neatly divided into two distinct and easily distinguished subgroups, descended from a common ancestor among the iguanodonts sometime in the Early Cretaceous period. Lambeosaurines, more robust in the hindquarters, with fewer teeth than hadrosaurines, are primarily distinguished by their crests. "Lambe's lizards" had, for the most part, high, elaborate head crests of hollow bone (at least on males). Hadrosaurines, like the long- and low-snouted *Maiasaura,* had flat heads with solid bony crests or humps. Nasal skin pouches on many may have inflated for display.

The crests differ markedly between duckbill species and between males, females, and young within a species. Their variability makes the job of identifying different duckbill genera relatively straightforward for us and for the duckbills themselves.

Flamboyant male crests, smaller female ones, and smaller-still juvenile crests did help the lambeosaurines know each other at a glance. Aside from recognition, the crests may also have served in courtship rituals, establishing rank in the herd, or in communicating from parent to offspring or herd member to herd member. The big nasal arch may have been used by battling males in smacking one another broadside, or pushing heads.

Arguably, what makes the duckbills as a group most special are their teeth, which make them, as Horner wrote, "the most sophisticated reptiles ever, living or extinct." Small teeth are packed closely together into columnar families, which are wedged together by the score (up to sixty families) upstairs and down in the duckbill jaw. In each tooth family, three to five replacement teeth crowd behind one to three working teeth in each tooth position. In all, as many as 1,200 teeth crowded one hadrosaur mouth.

Clearly these teeth were made for grinding. The plants duckbills devoured—weedy plants, shrubs, and needles and branches of conifers—were, for the most part, low-growing. So duckbills did most of their grazing around their feet, on forage less than six feet high. But they could comfortably crunch plants and fruits as high as thirteen feet off the ground.

They browsed in a world of rich, subtropical forests; small flowering trees were topped with a scattering of tall conifers. The fertile lower coastal plains of the North American West were dominated by swampy vegetation, ferns in the understory, conifers above. From its mummified remains, including possible stomach contents, we may have a good idea of what one duckbill ate—conifer needles and branches, deciduous foliage, and many little seeds or fruits.

For decades, paleontologists speculated on hadrosaur lifestyle with overly broad and flat wrong generalizations. The hadrosaurs' bills and on some, snorklelike crests suggested to paleontologists that all duckbills led a semiaquatic life. Those observations were in keeping with the swampy habitats envisioned for these and other dinosaurs. From the floodplain sediments of Dinosaur Provincial Park in Alberta (see Chapter 1), Dodson made the case two decades ago for duckbills as predominantly, but not exclusively, aquatic, perhaps like moose in daily use of watery environs.

Now the duckbills are seen as land-loving, capable of running on two legs, standing or ambling on all fours. With expanded excavations, particularly by Horner (and Dodson, Weishampel, and others as well), it is now apparent that certain duckbills preferred drier upland environments. Others dwelt in coastal lowlands.

Fossils of the broadest snouted, arch-faced hadrosaurs are found in what were deltas close to the ocean. Many of the crested lam-

David Weishampel of Johns Hopkins University is a scholarly paleontologist best known to the public for his studies of how duckbilled dinosaurs chewed and how some of them could have produced resonating sounds from their nasal tubes. [Photo by author]

beosaurines come from lower coastal lands inland of these deltas. Upland lived more conservative-appearing duckbills, such as *Maiasaura*.

Matthew Carrano, a precocious Brown University undergraduate, drew an intriguing parallel between duckbills and modern cud-chewing mammals in a presentation at the 1991 SVP meetings in San Diego. Carrano noted that among today's ungulate mammals, those that showed marked differences in appearance between sexes also seemed to have shorter legs, narrower muzzles, and a more selective diet. These territorial browsers live alone or in small groups in more closed, forested habitats than the larger, less sexually distinct herding grazers of the more open savannahs.

Carrano's statistical studies of hadrosaur fossils showed a similar relationship between size, tooth wear, and sexual dimorphism from crestless duckbills to crested ones. The crested and sexually varied lambeosaurs may therefore have been less social and more territorial than the crestless hadrosaurines. And the crested duckbills should by Carrano's reckoning be found in lusher habitats than the crestless ones. What we know of duckbill environments does indeed support Carrano's assertions.

The Horned Dinosaurs

Life in the lands of the American West 75 million years ago was as good for the horned dinosaurs as it was for the duckbills.

Excavations of the Judith River Formation—the stretch of badland outcroppings from western Montana up to southern Alberta—have provided a wealth of horned dinosaur fossils, much as they did for duckbills. The riches are the abundant ceratopsian fossils; the embarrassment comes from naive and aggrandizing attempts to name them. Cope and other earlier dinosaur hunters of the American West engaged in a frenzy of horned-dinosaur naming.

Now, however, the ceratopsians are known well enough in all their horned forms to be organized into distinct groups. Paul Sereno, who reclassified them in 1986, identified more than a dozen derived characters that distinguished the group, among them a big head and a pointed rostral "beak." The beaks were used for grasping and plucking rather than for biting, and the close-packed teeth were for shearing tough, low-growing plants.

The horned dinosaurs belong to three families. One, the little parrot-beaked psittacosaurids that Sereno studied in Asia, aren't known from North America. They may well be the ancestors of all horned dinosaurs. The small and primitive protoceratopsids are known predominantly from drier and upland areas of eastern and central Asia as well as western North America. Some were quick bipeds. None were bigger than eight feet long. The familiar and more dramatic horned dinosaurs are the ceratopsids, among them *Triceratops*. The ceratopsids were midsize to large browsers by dinosaurian standards, seventeen to twenty-five feet in length as adults. They are known well only from North America, with fragmentary, and to many, dubious claims from South America and Asia (including most recently the Soviet finds from Kazakhstan).

And they are principally known from their skulls. Unlike the pattern for most dinosaurs, for ceratopsids skulls are relatively abundant, their postcranial remains rare. Not one complete skeleton of the familiar *Triceratops* is known.

Yet, these horned dinosaurs are the best known of any dinosaurs, from Alaska south through Alberta, Montana, and Texas to northern Mexico, and all from the last 15 million years of dinosaur life.

There they dominated the last known communities of dinosaurs, comprising one-half to seven-tenths of the total known dinosaur populations.

All the ceratopsids were stout, four-legged runners, rhinoceroslike in their bulk. Their leaf-shaped teeth were locked together in a complex series of a rows, dental batteries somewhat resembling those of duckbills.

They varied spectacularly, however, in the number and configuration of their nose and over-the-eye horns and in the bony frills extending from the back of their skulls. They offer an impressive array of headgear—some with long frills, others with short frills fixed with as many as six spikes. Ceratopsid hood ornaments include big brow horns, bony cheek growths, nose horns that grew big, short, upright, or drooping. Their particular cranial decor not only distinguished species, but males from females in a species. The presence of these identifying features common to many herding animals today strongly suggests, as do the single-species horned-dinosaur bone beds found by Currie and others, that these were gregarious, herding creatures.

It would not have been difficult for a female horned dinosaur to spot a male of its species on the Western dinosaur floodplain. Texas Tech University paleontologist Thomas Lehman documented sexual distinctions in the skulls from a population of a horned dinosaur from Texas. He suggested males had larger, straighter, more erect and vertical eye horns than females of that species. An analogy could be drawn to many elaborately horned male ungulates on the Serengeti today, or to male killer whales and their large dorsal fins.

And as Darren Tanke, a Royal Tyrrell Museum paleontologist, documented from a centrosaur bone bed, young dinosaurs show slow growth of head armor. By delaying their acquisition of adult head gear, young male horned dinosaurs could have avoided becoming contestants in courtship and dominance contests until they were fully mature.

The adult male's head equipment was well suited to contesting openly for females and for dominance in this environment. Dodson and Currie imagine two horned-dinosaur "bulls approaching each other with their heads lowered and their crests elevated in a magnificent and colorful display. If bluffing failed to establish dominance, then the orbital horns were ideally suited for locking together for a relatively safe pushing and wrestling match." And some horned dinosaurs might have "engaged in side-to-side combat, with the nasal horns and spikes on the frill locking as the animals pushed against one another."

At the 1991 SVP meetings, Scott Sampson, an enterprising doctoral student at the University of Toronto, presented a sweeping profile of horned dinosaur behavior. After analyzing the skulls of horned dinosaurs, Sampson concluded, as Dodson and Currie had, that the animals must have wrestled each other with interlocking horns as many horned ungulate mammals do today. Sampson noted that while the horned dinosaurs were alive, their horns would have been covered with hard sheaths, which probably had ridges to help grip their opponents' headgear.

The frills and horns of the horned dinosaurs appear to Sampson to be designed for intimidating and wrestling rivals rather than as weapons for use against predators. Farlow and Dodson interpreted the horns and frills of ceratopsians in 1975, noting that for the frill to be fully extended, the head would have to be lowered, nose to the ground. In this position the horns would not be oriented for attack. Yet they were dangerous. The puncture holes in the cheeks and frills of horned-dinosaur skulls suggest to Currie and Dodson as to earlier workers "that these animals fought with members of their own species." Heavily vascularized horns and frills also worked to control body heat.

With Horner and Weishampel's data the horned dinosaurs can be separated, like the duckbills, by habitat preferences. As with duckbills, some horned dinosaurs seem to prefer the lowlands and coasts, others favoring inland and high country.

But that is only part of the story of the evolution of horned and other dinosaurs in Late Cretaceous Montana.

All these dinosaurs and the shifting environment in which they lived are recorded in the sediments of the Judith River Formation. These strata also record in fine detail the movement of the Cretaceous inland sea. The changes in that sea are closely tied to changes in the dinosaurs of Montana. The challenge of discerning those patterns brought together a paleontologist from Montana and another from Ohio, putting Jack Horner and Dave Weishampel within one hundred miles and a few million years of each other in western Montana.

Comings and Goings

Weishampel is a tall and gentle man, still in his thirties, and already an accomplished scholar. From his rumpled corduroy jacket to his dark-rimmed glasses, to his fast, elliptical speech, he is as much the portrait of an Eastern academic as Horner is of a Western field man.

The son of a NASA worker, Weishampel grew up in suburban Cleveland. He went "crazy over dinosaurs" at seven and never fell out of love with them, though his interest became highly academic. For his master's research he chose to specialize on lambeosaurine anatomy "because it seemed suitably obscure." His research has produced many insights and one widely known conclusion, that duckbill crests could have blown low-frequency messages for miles, as trumpeting elephants communicate today.

Weishampel's doctoral work was a hefty thesis on the evolution of jaw mechanisms in duckbills and related dinosaurs. Weishampel did three-dimensional computer modeling of jaw systems in all these dinosaurs. He employed techniques borrowed from mechanical engineering and his own theories to make predictions about patterns of tooth wear and test them against actual tooth wear seen on fossil specimens.

He found, as British paleontologist Dave Norman had simultaenously and independently suggested, that duckbills and their kin developed a unique jaw movement. As they chewed, their jaws popped out to the sides slightly, sliding upper teeth over the coarse foliage, nuts, and berries sandwiched in their jaws.

This jaw motion was not only curious but also evolutionarily significant as it represents one of only three ways along which transverse chewing can evolve. So, says Weishampel, "dinosaurs can be taken out of their usual context and used to discover the evolutionary context" of a shift in how bodies function. Weishampel is still investigating duckbill jaw movement to determine which plants the motion might be best suited to chewing.

After completing his Ph.D., Weishampel took a fellowship in Tübingen, West Germany. There, he began indulging another of his interests, paleontological history, by reexamining the work of Franz Baron von Nopsca, an early twentieth-century Transylvanian paleontologist.

As Weishampel discovered, Nopsca's more unusual theories, and his sensationally bizarre life (a flamboyant manic-depressive homosexual, he attempted to make himself the king of Romania, and shot his Albanian lover and killed himself), had long denied him his proper scientific due. Significant to Weishampel and Horner's own studies, and to dinosaur science, were Nopsca's descriptions of his own backyard (literally, as his sister found dinosaur fossils on the family estate when Nopsca was a teenager)—the Siebenbürgen fauna of Romania—in one of the first full-fledged studies of dinosaur paleoecology. Late in his career, Nopsca suggested Siebenbürgen was an isolated Late Cretaceous island. Isolation would explain much about these odd dinosaurs.

Since the dinosaurs of Romania were among the last of their kinds, one might expect them to be the most advanced in form. On the contrary, the Transylvanian dinosaurs—among them, a pig-snouted plant-eater, a stout duckbill, and a sauropod that may have had horn pedicles like those of a giraffe—were peculiarly small and primitive, and less diverse than mainland dinosaurs of the time. "The hadrosaur is one of the last of the hadrosaurs, and it is the most primitive," says Weishampel.

Why? Diminished body size is common on islands. The founding population might have been of small dinosaurs, the limited resources available might have favored the survival of smaller animals, and isolation may have insulated that population from the effects of competition with more advanced invading dinosaurs. Isolation would also have isolated these dinosaurs from genetic change. Says Weishampel, "Evolution on an island is a bit like going for a ride with no return ticket. You're stuck the way you are."

The Romanian dinosaurs constituted a relict group and might significantly validate modern evolutionary theory of "island effects" on speciation postulated by biologists Ernst Mayr, and E. O. Wilson, and Robert MacArthur. Weishampel was eager to examine Nopsca's findings and subsequent Romanian discoveries and those elsewhere in Europe in that theoretical light. "Next to the fossils in North America and Asia, Transylvanian dinosaurs are the best we've got from the Late Cretaceous of the northern hemisphere. Now that we know more about the geology and paleobiology of the time, we ought to say more about the relationship of these dinosaurs, how they got there, who they are related to, and put Nopsca's island biogeography into proper focus."

In 1985, Weishampel joined forces with Romanian paleontologist Dan Grigorescu, and Oxford University paleontologist David Norman, who had long been interested in Nopsca's dinosaurs, to study the Nopsca material in the British Museum of Natural History. In 1978, half a century after Nopsca's work on dinosaurs, Grigorescu had begun his own dinosaur excavations in the Hateg Basin of the Transylvanian Depression where Nopsca had found dinosaurs, in hopes of restoring Nopsca's collection. Grigorescu has now collected more than four hundred isolated bones from seven different dinosaur taxa, and several dinosaur eggs.

From their studies of Nopsca's dinosaurs in London, Bucharest, and elsewhere in Europe, the team has a good sense of the animals' anatomy and evolutionary status. Of the hadrosaur, Weishampel says, "It shows a lot of iguanodont characters. This animal's dentition is not Joe Hadrosaur's, it's much more primitive." The hypsilophodont (Nopsca thought it was an iguanodont) is, according to Weishampel, "very peculiar. Hypsilophodonts are lightly built. This one is a tank." Though it was less thoroughly studied, says Weishampel, "the sauropod is really small, only twelve to eighteen feet."

As Weishampel now sees it, the Transylvanian animals are indeed small and primitive island relics, just as Nopsca suggested a half century ago. Environmental events, but of a very different nature, also shaped the evolution of the duckbills and horned dinosaurs Weishampel and Horner are unearthing in Montana.

Comings and Goings, Part II

Horner and Weishampel are working on a parallel thread of Late Cretaceous dinosaur evolution. They are following the waxing and waning of a vast inland sea to discern which species exploit these habitats and which expire along the shifting margins. Unlike the sheltered retrogrades of Transylvania, the Montanan dinosaurs were confronted and sometimes eliminated by the movement of the sea around them.

Weishampel and Horner met when Horner was still at Princeton and Weishampel was a Philadelphia graduate student. "We were both self-styled hadrosaur workers," says Weishampel. When Horner had found more Montanan dinosaur sites than he could investigate, he invited Weishampel to take over those from the St. Mary's Formation, several hundred thousand years more recent in dinosaur time than the age of the bulk of Horner's study fossils in the Two Medicine Formation.

The Two Medicine is two thousand feet deep, recording a time span from 84 million to 72 million years ago. Those were changing times for Montana and its dinosaurs. The Late Cretaceous dinosaurs worldwide witnessed flooding unsurpassed since, and for half a billion years before them. Sea levels had been on the rise, with a few dips and peaks, through the Jurassic, producing vast shallow seas, but nothing to equal the Late Cretaceous high water marks. Less than 60 percent of present-day land on earth was above water. At their highest, the Cretaceous seas swelled to more than six hundred feet above present sea level. (The same sea rise would submerge only half as much land today, as mountain building has made modern lands higher and hillier than in dinosaur days.)

The Cretaceous flood was not one event but at least five worldwide, each lasting 2 to 4 million years. In North America, the incursions of the sea created the shallow inland sea. That Colorado sea washed across Montana as it spread west from the Rockies to the eastern lowlands and south from the Arctic to the Gulf of Mexico. The Montanan shores, laden with sediments sloughed from the rising Rockies, were fertile lagoons, swamps, and deltas—a fine place for dinosaurs and other living things, while it lasted. How long the high water lasted, and what replaced it, is recorded in the banded badlands of western Montana. Among these strata are the layers of the Two Medicine Formation.

"What the Two Medicine Formation preserves, then, is the record of 12 million years of life on a coastal plain," writes Horner. As the sea swelled, it spread three hundred miles or more in less than a million years, or about two inches a year by Weishampel's estimate. It moved toward the mountains, inundating the land, and over many generations pushed the dinosaurs into an ever-smaller habitat. This time of what Weishampel refers to as "species-packing" ought, he says, "to increase extinction rates, obviously, but at the same time up the speciation rates as well."

This evolutionary logic cuts, at least superficially, across common sense. Crowding cuts down on the number of animals, eliminating some species entirely. That's a familiar Darwinian scenario—survival of the fittest. But the survivors of this packing, the victors in the competition, are stressed and small in number. In this limited gene pool, new and unusual forms express themselves throughout the population more quickly. New species are made as others die off.

Before the sea moved in at the beginning of the time recorded in the Two Medicine Formation, iguanodonts predominate among the area's dinosaur fauna. After the sea receded, more than 80 million years ago, hadrosaurs, specifically an ancestor of the conservative-looking *Maiasaura* and a "hook-nose" genus Horner has yet to name, took their place. The proto-*Maiasaura* appears to have hugged the upper plains, while the "hook-nosed" opportunists radiated quickly in the coastal plain. So, too, with horned dinosaurs, Horner believes. Droopy-nosed styracosaurs acted as opportunists, while more conservative, less ornately decorated horned dinosaurs persisted in the highlands. Horner found that this pattern of settlement persisted for more than 10 million years thereafter as the sea moved in and out, altering the environment and the dinosaur populations as it went.

These days, Horner moves from site to site at different levels within the Two Medicine Formation, following its changing sediments, the ancient landscape and the movement of the inland seas over millions of years. "We're trying to collect ten or fifteen individuals from a species from each period," he says. This ambitious plan presupposes a wealth of dinosaurs and a skill at finding and identifying them unheard of before Horner's explorations.

All these finds appear to fit well within the predicted evolutionary pattern of dinosaur species formation in response to the movements of the inland sea. In good times the elaborately crested and horned dinosaurs flourished in the lowlands. As the sea crowded in on the mountains, the conservative dinosaurs survived and new forms evolved, flowering as the seas receded once again.

But elsewhere along the ancient shore the fossil evidence does not so nearly support the Horner and Weishampel theory of conservative upland dinosaurs and specialized delta dinosaurs. In 1986, in his ongoing digs in the St. Mary's Formation of western Montana, only slightly younger than the youngest Two Medicine Formation sediments, Weishampel obtained puzzling fossils. The animal was *Montanoceratops,* "a really cute little guy, four feet long." Though it is supposedly the most derived of all proto-horned-dinosaurs, Weishampel found it in older rock than its less advanced cousin, *Leptoceratops.* "You'd expect the more primitive animal, *Leptoceratops,* to be the one that lived upland," Weishampel admits. "But the first colonizer is *Montanoceratops.*"

Weishampel concludes that the digs so far "may reflect how little we know. We need better sampling, to understand the ceratopsians and their spin-offs better."

Horner can't, or won't, explain the discrepancies either. Nor can Horner explain the ever-growing public fascination with dinosaurs, any more than he can his own lifelong one. "All I know is it's fine with me. The more people who love duckbills, the better."

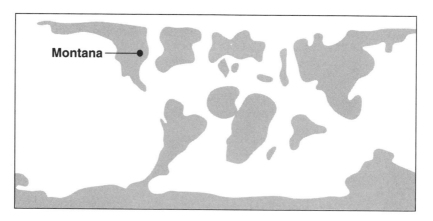

The breakup of the supercontinents, 65–64m. yrs. ago

T   J   C

65-64m. yrs. ago

245m.   65m.

AGE OF THE DINOSAURS

Triassic   Jurassic   Cretaceous

First animals   First land animals   First human

T   J   C

590m.   350m.   245m.   208m.   145m.   65m.   Today

CHAPTER 11

Dinosaurs at the Brink— and Beyond?

THE BANG GANG

What happened? What happened?
The dinosaur's extinct!
What happened? What happened?
Let's really try to think.

Too cold? Too hot?
Some changes in the weather?
No sun? No food?
A lack of fur or feather?

What happened? What happened?
A theory comes and goes.
What happened? What happened?
The truth is, no one knows.

Reprinted from *Wee Sing Dinosaurs,* copyright 1988, with permission from Price, Stern, Sloan, Inc., Los Angeles, California. Words by Susan Nipp.

That song, a favorite of my seven-year-old, sums up quite neatly the dinosaur extinction debate. The shape of the debate is, of course, ever changing. Theories are spun off at a dizzying rate, as the search for the mechanisms of extinction goes on. All go into far greater length than these lyrics, many with no greater certainty to their speculations. The following condensations of current extinction theses fall somewhere in length between the song and the many book-length themes. But like the long and the short of dinosaur extinction theories, they offer only tantalizing scenarios. No, Virginia, there is no certain, single cause for the dinosaurs' demise. Possibly, climatic change, volcanic eruptions, *and* a disaster from space did in the dinosaurs.

As hot-bloodedness was the dinosaur cause célèbre of the 1970s, sudden extinction was the vogue of the 1980s. Hot-bloodedness was largely imposed upon mainstream dinosaur paleontologists by the research and lobbying of maverick paleontologist Bob Bakker. Sudden extinction came from even farther outside, from autocratic physicist Luis Alvarez.

"Paleontologists tend to be somewhat defensive," Stephen Jay Gould says. "They suffer under the prejudice that theirs is a B science. And then some Nobel Prize–winning physicist tells them something." Bakker mocks the faddish comic-book appeal of sudden dinosaur extinction: "An iridium bullet in the braincase! Who was that masked comet!"

Dinosaur paleontologists are reluctant to hop on the sudden-extinction bandwagon for sound intellectual reasons. Just as the present body of evidence does not support universal dinosaur warm-bloodedness on the mammalian level, the fossil record does not argue conclusively for dinosaur extinction by extraterrestrial causes alone. "All was not well in the Late Cretaceous," says Gould, summing up prevailing paleontological opinion, not just his own. "It was not a good time in general for life. Otherwise, the impact would have been nothing more than a tiny blip in extinctions."

On that point the sentiment of dinosaur researchers is nearly unanimous, no matter who is asked. The idea of an extraterrestrially caused dinosaur extinction is "about as dead as the Cretaceous," according to paleontologist Kevin Padian. How else, then, do these scientists account for dinosaur extinction? Chinese paleontologist Dong Zhiming summed their view: "Many theories exist, but answers are few. Most likely it will involve multiple factors rather than a single catastrophic event."

Since the early 1900s, when Cretaceous and Tertiary sediments in the American West began to be explored in detail, scientists have known there was an extinction, one that appeared to mark the end of the dinosaurs, at the close of the Cretaceous. And ever since, they have been attempting to account for the demise of the dinosaurs with causes from constipation to poisoning to stupidity, which often read as ludicrous speculation today.

Environmental changes less dramatic but equally invidious as those produced by a meteoritic impact have also been invoked in the past to explain extinction. Reversal of the earth's magnetic field, a relatively common, if mysterious, phenomenon in earth history, was thought by some to have left the earth in flux at the end of the Cretaceous and susceptible to dinosaur-killing radiation.

Volcanoes enjoy continuing popularity as dinosaur murder weapons. Indeed, evidence exists for massive, if nonexplosive, eruptions in the Deccan Traps of India, basalt laid down miles thick in the Late Cretaceous and now preserved as the 200,000-square-mile Deccan Plateau—the largest known flow deposits on earth. The modus operandi followed by the killer volcanoes included, in theory, destruction of the ozone layer by hydrochloric acid from volcanic gases, carbon dioxide saturation of the atmosphere, and a resulting "greenhouse effect."

There have been climate-induced-extinction scenarios as varied as the estimates of the weather conditions themselves, or the Three Bears' porridge. It got too hot, as from carbon dioxide buildup. Or too cool, and as isolated arctic waters suddenly reunited with other oceans by continental movements, pouring into the warmer Atlantic, world ocean temperatures plummeted, causing a worldwide chill and drought. Or it was just right, and dinosaurs overpopulated and died of stress.

Imaginative researchers are still spinning off enticing and entirely unverifiable new scenarios. Bakker's notion of intercontinental migrations and ensuing epidemics is one. Several paleontologists have suggested that a sharp temperature rise may have caused the dinosaurs to lay eggs of only one sex, as reptiles respond today to warming. The dinosaurs would have died out then for lack of mates.

Love That Bomb

All these exotic scenarios were possible, none probable. And none has captured public attention like the asteroid hypothesis. University of California/Berkeley geologist Walter Alvarez, accompanied by a team from the University of California and Lawrence Berkeley Laboratory, found a large enrichment of the element iridium in a pencil-thin, 66-million-year-old layer of rock from Gubbio, Italy, in the spring of 1977.

The earth's supply of iridium is concentrated at the planet's core. Other concentrations, "spikes" in measurement readings, come from some forms of extraterrestrial objects, whether comets, meteors, or asteroids, that collide with the earth. Alvarez, his Nobel-laureate physicist father, Luis, and Berkeley nuclear chemists Helen Michel and Frank Asaro suggested in 1980 that the end-of-Cretaceous extinction was caused by the impact of an asteroid six miles wide. This event alone was both necessary and sufficient to account for the end of the dinosaurs, according to the authors. "Dinosaurs did last for nearly 140 million years, and we believe that had it not been for the asteroid impact, they would still be the dominant creatures on earth," the Alvarezes wrote.

The thesis was not new. For a decade before the Alvarezes proposed their theory, prominent scientists had been attributing mass extinctions to various extraterrestrial events.

The Alvarezes, however, had more evidence—the freshly obtained iridium spikes—to support their stand. Nonetheless, initial scientific reaction was far from universally supportive. "Most of my colleagues were just scandalized," Walter Alvarez recalled to reporters.

However, substantiating evidence soon began appearing of many more iridium spikes at the end of the Cretaceous around the world. Geochemical, botanical, and arguably, paleontological evidence continues to mount in favor of some version of an end-of-Cretaceous extraterrestrial impact. This evidence has convinced many scientists of various disciplines, but not all, that such an impact or impacts had some effect on the end of dinosaur life. With the public, however, the thesis of a killer asteroid has gained widespread and nearly unqualified acceptance, having occasioned media attention such as few scientific theses have ever received. Merits of the mass extinction hypothesis aside, it was, for the public, an easy-to-grasp solution to a topic of long-standing public interest: "an elegant and parsimonious solution to a question firmly embedded in popular culture, at least in the United States," wrote sociologist Elisabeth Clemens (daughter of gradual-extinction advocate and University of California/Berkeley paleontologist William Clemens.)

To counter this theory, paleontologists had only more complicated, antiquated, and sometimes conflicting accounts of a gradual extinction, and few spokespersons as forceful as the sudden extinction advocates, who, in addition to the Alvarezes, included Carl Sagan and Stephen Jay Gould. The disproportionate attention given in the popular and scientific press to the sudden extinction hypothesis was therefore not surprising.

The new ideas that emerged from the sudden extinction hypothesis have subjected the thesis to considerable refinement, in hundreds of publications, in a decade's time. One key piece of evidence, a "smoking gun," that was long missing from the asteroidal impact extinction theory was an impact crater of the right age and dimensions to account for a worldwide catastrophe—65 million years old and one hundred miles wide. Among the 120 well-documented craters on earth, none is larger than ninety miles across. However, asteroids that strike the seafloor or fall in the most remote locales are seldom detected.

Promising craters from a dinosaur-killing object have lately been found. In 1990, Soviet scientists produced a sample of glassy rock from the center of crater near the town of Popigay in Siberia, north of the arctic circle. The crater has been dated to 66.3 million years, close indeed to the theorized end of the dinosaur era.

Alan Hildebrand, a graduate student at the Lunar Planetary Laboratory of the University of Arizona, concluded in the late 1980's that something crashed into the Caribbean east of Nicaragua at the K/T boundary, a conclusion based on what he interpreted as deposits left by a tsunami (a long-period wave caused by seaquake or eruption and popularly, if inaccurately, known as a tidal wave) along the shores of the Caribbean and the east coast of the U.S., south from New Jersey.

Hildebrand's 1989 analysis of rocks found in the mountains of Haiti in 1975 by Haitian researcher Florentine Maurrasse offered support to a Caribbean-impact scenario. Geologists discovered tektites, tiny droplets of glass produced only under extreme heat. The glass bore neither gas nor water, telltale indicators of production by volcano. And chemical tests indicated that the tektites were 66 million years old, from very near the time of the K/T boundary. The relatively large size of the tektites and their dense concentration suggests that their Haitian source was close to the epicenter of the impact.

The possible site of that impact had been identified in 1978 by Glen Penfield, a Texas petroleum geologist whose aerial surveys of the Yucatan showed huge rings of abnormal magnetic readings from below the surface of the Yucatan in southern Mexico. Penfield's findings appeared only in a single paragraph of a scientific paper in 1982. Hildebrand located the paper and Penfield. Penfield began his own search for core samples from wells drilled within the possible impact area. Iridium or shocked quartz within the samples would support his and Hildebrand's contention that this was the site of the K/T boundary asteroid impact.

But a fire had destroyed most of the original Pemex oil company core samples. Penfield did succeed in locating samples at the University of New Orleans, and in 1990 Hildebrand and his colleagues analyzed the rock sample. They found limestone of extreme uniformity for 10,-000 feet, except at a depth of 1,000 feet where jumbled and broken green rock interrupted the limestone column. Hildebrand and his colleagues discovered shocked quartz within the odd sample, quartz that might well have formed when a crater was made nearby. The age of the rocks is still being tested, but preliminary results suggest 65 million years ago, the long-held date of the K/T boundary.

In 1991 more supporting evidence surfaced for a Yucatan K/T

boundary crater. Scientists at NASA's Ames Research Center detected a semicircle of sinkholes in the limestone that might have resulted from a slump in the deep underground rim of the buried crater. (Only half the crater rim is detectable as the rest is under water.) And in the fall of 1991, University of California/Berkeley postdoctoral student Nicola Swinburn announced to the Geological Society of America annual meeting that she and her colleagues had found tektite glass along the Mexican coast and coarse sediments nearby that indicated an "enormous physical disturbance of the seabed."

In sum, the most celebrated scientific detective quest of the decade has now produced a "smoking gun." The presumed dinosaur-killer is a crater 120 miles wide formed by an object from space six miles across slamming into the earth with an impact 10,000 times more powerful than the explosion that would be produced by setting off all the world's atomic weapons simultaneously.

The crater has been dubbed Chicxulub, for the town at its hub. In ancient Mayan, Chicxulub means "tail of the devil."

More Weapons

What sort of object struck the Earth 65 million years ago is less clear to the physicists investigating the extinction question. A meteor, a crater, or a comet might have left such a mark.

Since the Alvarezes's hypothesis appeared, clearer indications have been obtained of the frequency with which asteroids and comets crash into the earth. Gene Shoemaker of the USGS Planetary Division in Flagstaff, Arizona, and his wife, Caroline, have been searching since 1982 for asteroids that cross the earth's orbit. They've expanded the list of such objects from thirty to more than eighty in that time and predict there may have been hundreds more. Shoemaker has calculated that it is likely an asteroid more than six-tenths of a mile in diameter will hit the earth once every 40 million years. One that came very close, by astronomical distances, was reported on April 19, 1988. It passed within five hundred thousand miles of the earth the previous month, the nearest miss for an asteroid in fifty years.

University of Chicago paleontologists David Raup and Jack Sepkoski, accepting the sudden-impact-caused dinosaur extinction, argued, from marine extinction evidence, that mass extinctions have occurred with 28-million-year regularity over the last 250 million years. The cause of these periodic extinctions was suggested by Luis Alvarez protégé Richard Muller at Lawrence Berkeley Laboratory. Comets, though largely composed of ice, contain iridium and could be the source of the iridium found by the Alvarezes on the earth's surface. Muller suggested that our sun has a dark companion star, with an orbit of 28.4 million years. When that dark star—Nemesis, Muller called it—approaches the sun, it shakes loose a shower of comets from the Oort Cloud, a comet sea that lies beyond the orbit of Neptune. Muller hadn't found Nemesis in 1984 when he made his announcement, but predicted he would do so in six months. (Eight years later he has yet to discover Nemesis.) An alternative hypothesis holds that a Planet X, with a sharply inclined and shifting elliptical orbit outside the known planets, intersects the Oort Cloud every 28 million years. Yet a third explanation cites the bobbing motion of the solar system as it revolves around the center of the Milky Way. As the solar system passes through the middle of the galaxy, the gravitational pull of clouds of gas and dust concentrated there might tug comets into the solar system.

Shoemaker now thinks that a giant comet passing very close to the sun might have spewed out fragments that struck the Earth, a large one creating the Chicxulub crater, a smaller one making the nearly contemporaneous mile-wide Manson crater in Iowa.

After the Bang

If such a cosmic blast did befall the dinosaurs and their environment 65 million years ago, what would have ensued? If Nuclear Winter scenarios hold, the skies would darken with the soot of the impact and perhaps of the raging fires and volcanic eruptions it might trigger. The darkened, polluted earth would be gripped by drastic shifts in temperature and precipitation. And all this horror should have left its mark.

In 1986, Ronald Prinn, an MIT chemist, gained immediate renown among his colleagues as "the sidewalk man" when he suggested at an American Geophysical Union meeting that acid rain, "strong enough to melt a sidewalk in fifteen minutes," as Clemens recalls Prinn's statement, had pelted the earth following the K/T boundary impact (though volcano buffs suggest that Deccan Trap eruptions might have caused the acid rain).

Ed Anders, a University of Chicago chemist and former Alvarez associate, offered evidence of another postimpact horror. He detected large quantities of carbon in the iridium-bearing clay layer in Denmark and New Zealand. This carbon, he stated, was soot from a worldwide conflagration caused by the asteroid impact.

The best evidence for a sudden mass extinction at the K/T boundary comes more from plants than from dinosaurs. To some experts it argues for a rapid and drastic wave of extinctions and a marked change in climate. To others the signs point to a more gradual and perhaps less drastic climatic change.

Paleobotanist Jack Wolfe believes that the transformation of the planet 65 million years ago was great, and rapid, by any standards. Wolfe believes one or more impacts created nearly immediate and long-lasting effects—dust and fire storms clouding the atmosphere and heating the atmosphere, then cooling it for a far longer period as dust blocked sunlight. Wolfe's studies, some of the latest and most detailed analyses of the Late Cretaceous and Early Tertiary climate, go further still, suggesting with astounding specificity when the meteorite struck—on a day in early June.

Wolfe announced his calculations in 1991 based upon his studies of fossilized plants from what was a lily pond in Wyoming 66 million years ago. Some of the plants had blossomed and others had not yet bloomed, suggesting that the fossils recorded conditions in midspring. Yet all were shriveled, damage Wolfe attributed to a sudden mass freezing. To Wolfe, the unseasonable cold could only have been caused by the aftermath of an asteroid impact in which clouds of soot would have blocked sunlight from reaching the ground.

If the impact occurred in limestone (like that of the Yucatan where an object from space is thought to have landed), carbon dioxide released from the rock by the blast would slowly fill the air after the soot settled, producing a greenhouse effect. And that long warming, Wolfe says, is recorded in the fossils of plants with larger leaves and long "drip tips" characteristic of warmer, wetter climates than those of late-dinosaur days.

Wolfe contends that the impact and its fallout brought a sudden end to a long period of Cretaceous climatic stability. Climatic conditions in the Late Cretaceous were warm and near constant year-round right to the end, in his view, especially in lower latitudes. He asserts that drought or freezing conditions would be recorded in growth rings and air bubbles in the wood of Late Cretaceous trees. But Wolfe sees neither in the fossil wood he's examined. Rather, he says, the predominant vegetation remained green, year-round. "It warmed up a couple of, maybe three degrees in the mid-Maastrichtian [the Maastrichtian, 74 million to 65 million years ago, is the last accepted time for dinosaurs], but we have no change for a couple of hundred thousand years up to the boundary," says Wolfe. (He is speaking of middle- and low-latitude environments only.)

Suddenly, however, after the boundary, the vegetation shifts to a far higher diversity of leaf-dropping plants. Deciduousness, Wolfe asserted in 1987, was favored by the "impact winter" occasioned by the collision. The aftermath of the impact was not drought but a marked increase in rainfall. The world was dark and wet, and soon densely forested, after the boundary. This new world would not have been a conducive place for dinosaurs, if any had made it past the boundary, according to Wolfe. "A closed forest is not a good place for large herbivores."

In the Late Cretaceous a trend began, one that accelerated following the boundary, toward a higher diversity of smaller plant-eaters. The little vegetarians' victory may have been facilitated by the spread of denser, closed-canopy forests. The evidence of this moister climate is, Wolfe wrote, the thick coal deposits of the Early Tertiary. The drip tips on Tertiary leaves also suggest the weather had shifted from subhumid before the boundary to humid after it.

Take away the big plant-eaters, as the K/T extinction did, open up the terrain by fire, as a "Nuclear Winter" impact aftermath might have done, and staunch the fires by heavy acid rains and what remains? A highly competitive situation among plants seeking to establish themselves in the new environment, however stressful that environment was. What Wolfe calls the "throwaway leaf" would be of great advantage to plants colonizing the devastated planet, and coping with low or nonexistent winter light. In the first million years after the end of the Cretaceous, northern-hemisphere vegetation was impoverished in variety and dominated by ferns, consistent with Wolfe's scenario.

The impact might well have been limited to the northern hemisphere in its drastic environmental effects. Deciduous plants remain rare in the southern hemisphere, according to Wolfe, following the end of dinosaur time, suggesting an impact had little effect on southern-hemisphere vegetation and climate.

It might also have been cold in the months following the impact. Soon, however, and for many thousands of years thereafter, "it was warm, very warm," says Wolfe. From his analysis of New Mexico leaf fossils just after the boundary, says Wolfe, "the leaf shapes, margins, and size all tell us it was ten degrees Celsius warmer." In a geologically short time, the weather went from what he notes, from Latest Cretaceous eastern Wyoming, was "a mean of sixteen degrees Celsius, like Georgia today," to twenty-six degrees Celsius at his postboundary sites—"a full-blown tropical climate." Perhaps it got too hot, as well as too wet and overgrown, for the dinosaurs.

Wolfe's analysis of a deadly day in June for dinosaurs has been strongly criticized by paleobotanists who have studied the floras of the last dinosaur environments in greater detail. "Hogwash," says Leo Hickey, a Yale University paleobotanist, of Wolfe's scenario. Hickey and his former student Kirk Johnson (now of the Denver Museum of Natural History) analyzed thousands of leaves from before and after the K/T boundary. They see no evidence of freezing and account for the shriveling of leaves Wolfe called frost-damaged as simple decay. They also assert Wolfe misidentified several plants and lacked sufficient evidence for his estimates of shifting temperatures.

Nonetheless, from plants he's collected in the last ten years, Hickey has changed his thinking on dinosaur extinctions. Whereas once he believed in a gradual climatic degradation leading to mass extinctions, now he sees relatively sudden catastrophic climate changes close to and after the K/T boundary that he can only account for by an impact of an object from space. A shift to fern-dominated plant communities indicates to Hickey that conditions became radically warmer, perhaps several degrees at the K/T boundary. But unlike Wolfe, Hickey defines sudden change not as a month but as 250,000 years. The fossil record does not, he asserts, allow for more precise estimates of change than a quarter million years.

Other paleobotanists in the western United States and East Asia have, like Hickey, noted another "spike" in the record at the K/T boundary aside from the sudden increase in iridium. This spike marks the odd and short-lived abundance of ferns just above the K/T boundary in New Mexico, Montana, Saskatchewan, and Japan. The fern spike may well record a marked disruption of the plant environment, followed by a recolonization by ferns. The same fern opportunism following the 1982 eruption of the El Chichon volcano in Mexico. Dramatic as it is, the fern spike reading is to date a western U.S. and East Asian phenomenon, not a worldwide event.

Many plants did die off at the boundary of dinosaur life. Pollen experts have observed a sudden extinction of nearly half of all fossil pollen-producing plants at the boundary—a huge and sudden drop-off of a fifth to a half of their numbers at K/T boundary sites in Montana.

But to other paleobotanists, there was already a decline in plant diversity and climatic equability before the boundary. Robert Spicer and paleobotanical colleague Judith Parrish had found something different from what Wolfe discovered in their high-latitude studies of the Late Cretaceous environment. The climate was deteriorating in Alaska long before the K/T boundary. The flora of Alaska at the very end of the Cretaceous was far less diverse than it had been earlier in the Late Cretaceous.

Where do these environmental analyses leave us? If there is any one answer that satisfies both Johnson's and Wolfe's reckonings of the environmental changes around the K/T boundary, it is that an extraterrestrial impact and its aftermath could have provided the finishing touches for a plant community already in flux.

But sudden extinction arguments of any form have failed to persuade many scientists who see a decline in diversity of flora and fauna before the boundary. Nor did it convince other scientists who thought the K/T picture was too fuzzy to see any pattern at all. Around the world there were gaps and noncorresponding dates in the extinction record.

"Ecosystems were decaying for at least two million years before the impact," paleontologist Steve Stanley of Johns Hopkins University told *National Geographic.* He favors long-term cooling as the cause of the environmental degradation. What caused that cooling is uncertain.

Responding to the many uncertainties over a single extinction-causing event, several scientists advocated a "step-wise" extinction at the K/T boundary. According to this step-wise scheme, species began to die out well before the boundary and continued to do so for a time after, a sequence some 3 million years in length.

The steps could be markers of volcanic activity or multiple impacts, according to its advocates. (Or immigration and disease, according to Bakker, also a step-wise extinction advocate.) The direct agents of extinction might have been wholesale changes in ocean chemistry—a dip of two to five degrees, as indicated by oxygen isotope ratios in deep-ocean sediments. Organisms suited to tropical environments would be most cold-sensitive, and most affected by the shift in temperature. According to one step-wise extinction scenario, comet showers peppered the seas, causing tsunamis and bringing deep, low-oxygen waters in deadly contact with marine life. The ocean cataclysms might also have wreaked environmental havoc on land. However, no such step-wise shower of meteors has been detected. No evidence of multiple tsunamis has been detected, though Joanne Bourgeois, a Washington University student, announced in 1988 that she had discovered rubble at the iridium level of K/T boundary beds in Texas. This rubble bed, she stated, was created by a tsunami from an asteroid striking the end-of-Cretaceous ocean.

One of the latest step-wise twists on the asteroid extinction theme brings it into conjunction with the competing theory of relatively suddenly death, volcanic eruption.

At the December 1988 American Geophysical Union meetings, University of Rochester geologist Asish Basu put the two vying theories together with hard evidence, as others had done in theory earlier that year. Basu was summarizing the research he'd done on material sent him by Chatterjee and others. Basu's work, according to Chatterjee (see Chapter 3), might provide the "smoking gun" to settle the dinosaur extinction issue. "This was a step-wise extinction," says Chatterjee with characteristic excitement. "It was three, four, five million years of lava burping like hell, moving west, floor after floor, a million square miles—it continues into the ocean you know. Such large-scale volcanism the world has never seen!"

Chatterjee and colleagues had found dinosaur bones and egg clutches in the sandstone and limestone close up to the K/T boundary in Jabalpur, central India. And above them, in sandstone underlying the bottommost levels of the Deccan Trap lava flows, they removed sediment from which Basu discovered shocked quartz. To Basu, the discovery, "along with the recently established fact that the Deccan lavas erupted rather rapidly at the Cretaceous/Tertiary boundary, strongly indicates that the Deccan volcanism was the result of meteoritic impact." Since the volcanos themselves could not account for the shocked quartz, the explosive impact of a meteorite must have. And by blasting a huge crater that would lower pressure on the semimolten magma below the earth's crust, the impact would cause the lava to well upward, creating the Deccan Trap flows. Said Walter Alvarez, "If Basu has it right, it's a fine reconciliation between two viewpoints."

Whether Basu is right is uncertain. Both the dating and quartz structure at Basu's site have been questioned by critics. Basu admitted the evidence is, like that for all the arguments for sudden extinction of dinosaurs, only circumstantial.

Out With a Whimper

To many critics of either of the sudden catastrophe arguments, the evidence for a sudden mass extinction, or even any mass extinction, at the end of the Cretaceous is too shaky to support any killers-from-space speculation. Paleontologists, a mostly conservative lot, have been among the most vociferous and adamant of those critics.

For one, if the dinosaurs did die suddenly, paleontologists wouldn't know it. If dinosaur extinction was truly catastrophic, occurring in the lifetime of the longest-lived individuals, it happened far too fast to leave a mark in the rock record. The stratigraphic record on land defines periods closer to one hundred thousand years, at best, rather than a hundred, a generous guess for a dinosaur's life span.

Determining just what happened to animals at this time, in the sea or on land, has proven difficult because the K/T boundary era was not a time when sediments were deposited. Sea levels were dropping far and fast in the last few million years of dinosaur life, by three hundred to nearly five hundred feet. Not only would this obscure the rock record, it may likely have caused far-reaching environmental effects (more extreme weather on land, for one) and perhaps, extinctions in the ocean and on land.

When discussing the K/T boundary extinctions, some paleontologists have the dinosaurs specifically in mind. But the K/T boundary debate centers on a larger extinction question. Luis Alvarez said, "The problem is not what killed the dinosaurs but what killed almost all the life at the time."

The problem may also be that almost all life did not die at the K/T boundary. The K/T extinctions were far-reaching, though not nearly so devastating as some other extinctions (such as the one we are causing right now). In all, 85 percent of marine families and 86 percent of the nonmarine vertebrates survived the extinction. And many organisms on land and in the sea appear to have been dying off throughout the last few million years of dinosaur time.

Nondinosaurian reptiles in the water and the air—pterosaurs, mosasaurs, and plesiosaurs—disappeared. But on land, and along water margins, turtles, crocodiles, snakes, amphibians, and most mammals made it through the K/T boundary. Placental mammals and now-extinct multituberculates did fine, as did marsupials in South America, though they nearly went extinct in North America. Birds emerged in fine fettle. Perhaps they were all more adaptable than the dinosaurs to an impact aftermath, because they ate less, were better insulated, or could find shelter. However, paleontologist Clemens concluded otherwise in a thorough study, finding the extinction of animals at the K/T boundary was not closely correlated with body size, warm-blooded-ness, or cold-bloodedness.

Maybe the survival of other animals was simply a lucky break, of the kind that favored the dinosaurs' survival and spread more than 150 million years earlier. Whatever the reason, most land animals emerged from the K/T boundary almost unscathed. Their survival is a common element in the doubts of many paleontologists about the existence or extent of a catastrophic impact. Says Notre Dame University paleon-tologist Keith Rigby, Jr., "The fundamental flaw in the Alvarez hypothe-sis is how turtles and these other vertebrates most susceptible to tem-perature change came through it fine, mammals diversifying, and not dinosaurs."

In an interval of less than one hundred thousand years at or leading up to the K/T boundary, dinosaurs disappeared from the earth, unless one believes claims, shaky to date, by Rigby and others, that some pressed on into the Tertiary. Rigby thinks he has found dinosaur frag-ments above the K/T boundary, but fellow paleontologists question his interpretation. Until an articulated dinosaur skeleton is found in sedi-ments clearly after the boundary, we cannot assume any dinosaurs survived. However, if dinosaurs were extinguished at the K/T bound-ary, judging from their fossil record, they may not have disappeared with suddenness.

It is a curious accident of paleontology that dinosaurs are best known from their last days on earth. In the western United States and Canada, dinosaur fossils are relatively abundant from the last 15 million years of their existence. So, too, in Mongolia. They're also known from Mexico, Honduras, South America, Europe, the Middle East, Africa, India, Australasia, and of course, Alaska. In all, 40 percent of all known dinosaurs have been found from a time 80 million to 65 million years ago, less than 10 percent of their time on earth.

Though near-K/T-boundary dinosaur-fossil localities have been re-ported in India, China, South America, the Southwest, and arctic Can-ada, with but fragmentary bones at all those sites, there is, to date, only one place where these last of the dinosaurs have been studied in depth: the Hell Creek Formation of eastern Montana. There, dinosaur diversity

On the shores of the Ft. Peck Reservoir in eastern Montana, University of Notre Dame paleontologist Keith Rigby sifts rock for dinosaur bone he thinks dates from after the accepted end of dinosaur life. Many of his colleagues dispute Rigby's analysis of his finds.

declines, according to one survey, from nineteen genera at the bottom of the formation to twelve in the upper fifty feet and just seven at the very top.

This finding has lately been disputed by paleontologists Peter Sheehan of the Milwaukee Public Museum and Bruce Fastovsky of the

University of Rhode Island. In their census of myriad small dinosaur bones sampled from the Hell Creek of Montana, presented to the USGS annual conference in October 1991, they found no decline whatsoever in the number and diversity of dinosaurs present approaching the end of dinosaur time.

Being the only well-studied site in the world at the time of the end of the dinosaurs, the Hell Creek may or may not be representative. And whether or not it reflects most dinosaur habitats of the time, it is markedly different from river and floodplain sediments of Alberta where earlier Late Cretaceous dinosaurs are well known. The Hell Creek environment was a seasonally dry lowland, intercut with channels. The fossils it preserved didn't keep well. Few are intact. While a paleontologist can't predict what he or she will find next in the onetime floodplains of the Judith River Formation of Alberta, horned dinosaurs, dome-heads, or duckbills, in the Hell Creek it's a safe bet the new find will be a *Triceratops* (with a duckbill a distant second). If truly fortunate, one might find a *Tyrannosaurus rex*. There are only nine *T. rex* known, and most of those, including the two biggest and best, both excavated in 1990, come from the Hell Creek Formation.

What is apparent, at least from available North American evidence, is that, as Kevin Padian writes, "by the end of the Cretaceous, there are few kinds of dinosaurs left, apart from the large carnivores and the horned ceratopsians. . . . Long gone were the stegosaurs, and the hadrosaurs, ankylosaurs, pachycephalosaurs, and small carnivorous theropods were dwindling in numbers and diversity."

Perhaps some or all of these animals survived elsewhere, in wetter lowland habitats that haven't been discovered yet by paleontologists.

The question of a decline in dinosaur diversity before the boundary remains an open one in the mind of gradualist and catastrophist paleontologists alike. "There haven't been enough dinosaur studies done," says Clemens. "I'm not convinced the decline is not just an artifact of sampling," says Paul Olsen.

Dale Russell, a longtime loner among dinosaur paleontologists in favoring sudden dinosaur extinction, admits the same. "The data on dinosaurs is so crude. You can't use it to talk about extinction."

Out With a Whimper-Bang

It is, of course, entirely possible that gradual, step-wise and sudden extinctions all took place at the end of the Cretaceous. While not abandoning the case for gradual extinction, paleontologists have come to accept the distinct possibility, if not certainty, of a catastrophic impact.

Says University of Pennsylvania paleontologist Dodson, "Physical evidence for an iridium anomaly is so widespread at the K/T boundary that it cannot be ignored. The 'iridium event' has to be accounted for. I'm not denying the event may have happened. But I think there's support for saying dinosaurs were already severely stressed by changes in their environment."

What is certain about the greatest animals the world has known is that they are long gone. All that endures today are bits and pieces, and our fascination with their lives, and their deaths. We have made a rather arbitrary and sentimental choice in lavishing on them, among all extinct life, such attention. Early twentieth-century dinosaur collector Charles H. Sternberg wrote, "Animals come on the stage of life and exist for a greater or lesser period as it may happen and then disappear, and the old saw 'that every dog has his day' is literally true of the past as of the present."

Yet however the dinosaurs died, their death is cause for genuine sadness at our never having glimpsed them in their living majesty. And it is cause for reflection on our own good fortune, for without their passing, the world would not have made room for the accident of our evolution. Sternberg himself put it elegantly: "They, too, were wonderful, they are dead, and their death recalls to us something of the meaning of living."

Author's Note

For all its splendid recent achievements, dinosaur science is in a parlous state. Cutbacks in the already modest government and private support for so-called "soft sciences" jeopardize the research, and the jobs, of several dinosaur researchers in a field of but a few dozen scientists. The difficulties of making a living as a paleontologist have discouraged many promising and committed young researchers such as Jill Peterson (featured in Chapter 6), who found conditions so grim that she could make a better living as a schoolteacher.

Digging dinosaurs takes money. A single element of a dinosaur skeleton, a sauropod pelvis to take the most extreme example, can be six feet long, tons in weight. To dig it out, a crew has to be brought to a field site, likely in a godforsaken spot.

Reaching remote places is expensive. If and when paleontologists find something of scientific value in the short season that weather, finances, and university schedules allow, that something must be hauled to a laboratory. There, months, sometimes years, of labor by

[© Donna Braginetz 1988]

professional preparators are needed to free the fossil from its surrounding rock matrix. Most dinosaur researchers do their own preparation work in the winter, as few museums and universities can afford the paid help.

Lots of expenses, yes, but not huge sums. The total annual expenditure on dinosaur research in the United States and Canada combined is scarcely $600,000 by Peter Dodson's 1989 estimate. And that is quadruple the figure from a 1980 study by Canadian paleontologist Dale Russell.

That a realm of scientific endeavor should be so woefully underfunded is disgraceful, even if typical for many of the low-tech disciplines. Why do dinosaurs matter? As the most successful animals in the history of life on land, they offer particular insights into modes and pace of evolution. For the young they are a special case of fantasy turned safely real, and in that reality a uniquely appealing introduction to scientific inquiry. For me and many of us, dinosaurs and dinosaur explorations are a source of wonderment and pleasure. They are as important as any endeavor to understand the world as it once was.

But there is special, shameful irony about the lack of money for dinosaur research, for Dinomania is all around us. Dinosaurs have lately graced, to use the term lightly, toilet paper, ravioli, checkers, and innumerable T-shirts. They are the subjects of prime-time TV shows, documentaries, feature films, Saturday morning cartoons. They are made into posters, stuffed cuddlies, inflatable robots, even fitted with guns and monsters for boys' "action" toys.

Much of this dino-merchandise features inaccurate images of dinosaurs, such as the British Museum's plastic dinosaur replicas ("These are not toys" read their advertisements), which feature the sail-finned *Dimetrodon,* that creature of a far earlier age more nearly related to us than dinosaurs.

Unlike Ninja Turtles, dinosaurs aren't trademarked. They don't earn royalties and ancillary income from every tie-in that uses their name or their likeness. But those who made Dinomania possible, the scientists who supply the information that is used and abused so widely, could use a piece of the action and a chance to improve dinosaur products.

Along with many of the leading dinosaur scientists and artists, I've formed The Dinosaur Society, a nonprofit organization, to further dinosaur research and education.* The Dinosaur Society funds international research projects and provides a timely, accurate source of information in the form of regular publications for children and adult members of our organization. All of us are volunteering our services, and several are donating a portion of their incomes from commercial endeavors to this effort.

So, if you bought this book, you've already contributed in a small way to ensuring that there is a future in exploring the past.

If you would like more information about The Dinosaur Society and its programs, please write care of Post Office Box 171, Newton Lower Falls, Massachusetts 02162.

*Drs. Bakker, Dodson, Bonaparte, Currie, Dong Zhiming, Horner, Hotton, Kurzanov, Molnar, Norman, Osmolska, Sereno, and Weishampel, plus David Attenborough and Michael Crichton are among the board members.

Appendix A: Dinosaur Fossils and Footprints

The extent as well as duration of the dinosaur conquest of the earth remains poorly appreciated. Dinosaur remains have been found in every continent and thirty-one of the fifty U.S. states. Johns Hopkins University anatomist David Weishampel has documented those finds and listed them by time period in *The Dinosauria* (University of California Press, 1990). The following list of dinosaur finds by continent, country, region, and time period, and the corresponding maps on the endpapers of this book, are drawn from Weishampel's compilation and updated with his permission.

North America

E. Canada T, J
Western Canada C
Yukon C

Alabama C
Alaska J, C
Arizona T, J, C
Arkansas C
California C
Colorado T, J, C
Connecticut J
Delaware C
Georgia C
Idaho C
Kansas C
Maryland C
Massachusetts T, J
Mississippi C
Missouri C

Montana C
Nebraska C
Nevada C
New Jersey T, J, C
New Mexico T, J, C
New York T
North Carolina T, C
North Dakota C
Oklahoma J, C
Oregon C
Pennsylvania T
South Dakota J, C
Texas T, J, C
Utah T, J, C
Virginia T
Wyoming J, C

Mexico C

Europe

Austria C
Belgium T, C
Czechoslovakia C
England T, J, C
European USSR C
France T, J, C
Germany T, J, C
Hungary J
Italy T
The Netherlands C

Norway C
Poland T
Portugal J, C
Romania C
Scotland T, J
Spain J, C
Sweden J
Switzerland T, J
Wales T
Yugoslavia C

Asia

Afghanistan J
Central China J, C
Iran J
Israel C
Japan C
Kazakhstan C
Laos C
Mongolia C
Northeast China C
Northeast India J, C

Northwest China J, C
Siberia C
Southeast China C
Southeastern India T, J, C
South Korea C
Southwestern China T, J, C
Syria C
Thailand J, C
Western India C

South America

Argentina T, J, C
Bolivia C
Brazil T, J, C
Chile J, C

Colombia J, C
Peru C
Uruguay C

Africa

Algeria J, C
Egypt C
Kenya C
Lesotho T, J
Libya C
Madagascar J, C
Malawi J
Mali C

Morocco T, J, C
Namibia J
Niger J, C
South Africa T, J, C
Tanzania J
Tunisia C
Zimbabwe T, j

Australia

Northern Australia C
Queensland T, J, C

Southeast Australia C
Western Australia C

New Zealand

North Island C

Antarctica

J, C

The World Today

Dinosaur Fossils
& Footprints

T = Triassic
J = Jurassic
C = Cretaceous
▲ = All three

[Deborah Perugi]

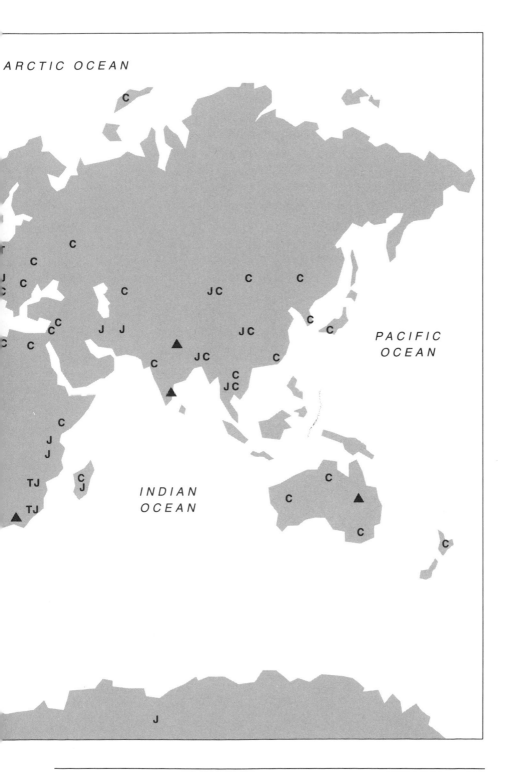

ARCTIC OCEAN

PACIFIC
OCEAN

INDIAN
OCEAN

Appendix B:
Where to See
and Dig Dinosaurs

If you'd like to see dinosaurs, visit one or more of these museums with major dinosaur collections and/or exhibits:

United States (alphabetical by name of institution)
Academy of Natural Science, Philadelphia
Alexander Ruthven Museums, University of Michigan, Ann Arbor
American Museum of Natural History, New York
California Academy of Sciences, San Francisco
Carnegie Museum of Natural History, Pittsburgh
Cleveland Museum of Natural History
Denver Museum of Natural History
Dinosaur National Monument, Jensen, Utah
Dinosaur State Park, Rocky Hill, Connecticut (you can make plaster-of-paris dinosaur-footprint impressions here)
Dinosaur Valley Museum, Grand Junction, Colorado
Dinosaur Valley State Park, Glen Rose, Texas
Earth Sciences Museum, Brigham Young University, Provo, Utah
Field Museum of Natural History, Chicago
Fort Worth Museum of Science, Texas

Geological Museum, University of Wyoming, Laramie
Ghost Ranch, Ruth Hall Paleontology Room, Abiquiu, New Mexico
Houston Museum of Natural Science, Houston
Los Angeles County Museum of Natural History
LSU Museum of Geoscience, Baton Rouge, Louisiana
Museum of Comparative Zoology, Cambridge, Massachusetts
Museum of Natural History, University of Kansas, Lawrence, Kansas
Museum of Northern Arizona, Flagstaff, Arizona
Museum of Paleontology, University of California, Berkeley, California
Museum of the Rockies, Montana State University, Bozeman
National Museum of Natural History, Smithsonian Institution, Washington, D.C.
Nebraska State Museum, University of Nebraska, Lincoln
New Mexico Museum of Natural History, Albuquerque
Peabody Museum of Natural History, Yale University, New Haven, Connecticut
Petrified Forest National Park, Arizona
Prehistoric Museum and Cleveland-Lloyd Dinosaur Quarry, Price, Utah
Science Museum of St. Paul, St. Paul, Minnesota
Teton Trail Village Museum, Choteau, Montana
Texan Memorial Museum, University of Texas, Austin
Trail Through Time, Grand Junction, Colorado
Utah Field House of Natural History State Park and Dinosaur Gardens, Vernal
Utah Museum of Natural History, Salt Lake City

International (alphabetical by country)

Museo Argentino de Ciencias Naturales, Argentina
Museum of La Plata University, La Plata, Argentina
Australian Museum, Sydney, Australia
Queensland Museum, Fortitude Valley, Australia
Natural History Museum, Vienna, Austria
Bernissart Museum, Belgium
Museu Nacional, Rio de Janiero, Brazil
Calgary Zoological Gardens, Alberta, Canada
Dinosaur Provincial Park, Patricia, Alberta, Canada
National Museum of Natural Sciences, Ottawa, Ontario, Canada
Provincial Museum of Alberta, Edmonton, Alberta, Canada
Royal Ontario Museum, Toronto, Ontario, Canada
Tyrrell Museum of Paleontology, Drumheller, Alberta, Canada
Beipei Museum, Szechwan, China

Chengdu Institute, Chengdu, China
Institute of Vertebrate Paleontology and Paleoanthropology, Beijing, China
Museum of Inner Mongolia, Hohehot, China
Museum of Natural History, Beijing, China
Qongqing Natural History Museum, Szechwan, China
Zigong Dinosaur Museum, Szechwan, China
Bavarian State Collection for Paleontology and Historical Geology, Munich, Germany
Geological and Paleontological Institute, Münster, Germany
Institute and Museum of Geology and Paleontology, Tübingen, Germany
Natural History Museum, Berlin, Germany
Senckenberg Nature Museum, Frankfurt, Germany
State Museum for Natural History, Ludwigsburg, Germany
Geology Studies Unit, Calcutta, India
Museo Civico di Storia Naturale di Venezia, Italy
State Museum, Ulan Bator, Mongolia
The Museum of Earth Science, Rabat, Morocco
Musée National du Niger, Niamey, Niger
Dinosaur Park, Chorzów, Silesia, Poland
Institute of Paleobiology, Warsaw, Poland
Bernard Price Institute of Paleontology, Johannesburg, South Africa
South African Museum, Cape Town, South Africa
Paleontological Museum, Uppsala, Sweden
Dinosaur Hall, National Zoo, Taipei, Taiwan
Birmingham Museum, Birmingham, United Kingdom
British Museum of Natural History, London, United Kingdom
Crystal Palace Park, Sydenham, London, United Kingdom
The Dinosaur Museum, Dorchester, Dorset, United Kingdom
Hunterian Museum, The University, Glasgow, United Kingdom
The Leicestershire Museums, Leicester, United Kingdom
Museum of Isle of Wight Geology, Sandown, United Kingdom
Royal Scottish Museum, Edinburgh, United Kingdom
Sedgwick Museum, Cambridge, United Kingdom
University Museum, Oxford, United Kingdom
Central Geological and Prospecting Museum, Leningrad, Russia
Paleontological Institute, Moscow, Russia
National Museum of Zimbabwe, Harare, Zimbabwe

A less complete listing, but including addresses and genera of significant fossils in each collection, is in Dr. David Norman's *The Illustrated Encyclopedia of Dinosaurs.*

Dinosaur Digs

If you'd like to assist in a paleontological exploration for dinosaurs, several institutions accept paying volunteers, from ages as young as five in some programs. They include:

The Dinosaur Society, Newton Lower Falls, Massachusetts

Boston Museum of Science, Boston, Massachusetts

Dinamation International Corporation, San Juan Capistrano, California

Earthwatch, Watertown, Massachusetts

Lawrence Hall of Science, University of California, Berkeley, California

Milwaukee Public Museum, Milwaukee, Wisconsin

Museum of the Rockies, Montana State University, Bozeman, Montana

Oregon Museum of Science and Industry, Portland, Oregon

Royal Tyrrell Museum of Paleontology, Drumheller, Alberta, Canada

Southwest Paleontology Foundation, Albuquerque, New Mexico

Appendix C: Recommended Reading

For dinosaur reading, several recent and accessible books make an excellent introduction to various aspects of dinosaur paleontology and those who practice it. Recommended are:

Alexander, R. McNeill. *Dynamics of Dinosaurs & Other Extinct Giants.* New York: Columbia University Press, 1989. A lively little book in which a scientist tackles problems of dinosaur locomotion.

Bakker, Robert T. *The Dinosaur Heresies.* New York: William Morrow, 1986. One man's view of dinosaurs, engagingly written and illustrated, and full of fascinating, controversial theories and speculations.

Booth, Jerry. *The Big Beast Book.* Boston: Little, Brown, 1988. There are a host of dinosaur books for the youngest readers. For those somewhat older with an enduring interest, or parents of younger dinophiles, I recommend this book with particular enthusiasm, as it is not only laced with accurate information, but provides a number of engaging experiments and activities to involve youngsters in understanding dinosaurs, paleontology, and evolution.

Czerkas, Sylvia J., and Everett C. Olsen. *Dinosaurs Past and Present*. Vols. I and II. Los Angeles: Los Angeles County Museum of Natural History, 1987. Two hugely oversize and magnificently illustrated paperbacks. They present summaries of recent research by several controversial paleontologists in their own words, accompanied by the best in dinosaur illustration, historic and contemporary.

Czerkas, Sylvia and Stephen. *Dinosaurs: A Global View*. New York: Bantam Doubleday Dell, 1991. A huge and beautifully illustrated new overview of Mesozoic life.

The Diagram Group. *The Dinosaur Data Book*. New York: Avon Books, 1990. A handy paperback reference.

Horner, John R., and James Gorman. *Digging Dinosaurs*. New York: Workman Publishers, 1988. A concise and very readable description of how one uniquely talented paleontologist looks for and analyzes dinosaur fossils.

McGowan, Christopher. *Dinosaurs, Spitfires & Seadragons*. Cambridge: Harvard University Press, 1991. A scientist looks at the lifestyles of dinosaurs and other animals with insight and style.

Norman, David. *The Illustrated Encyclopedia of Dinosaurs*. New York: Crescent, 1985. The best popular compendium of dinosaur information yet published, prepared by a leading paleontologist and lavishly illustrated by John Sibbick.

Preiss, Byron, ed. *The Ultimate Dinosaur*. New York: Bantam Doubleday Dell, in press, 1992. A well-illustrated, chronological compendium of essays by leading dinosaur scientists and science-fiction writers, and me.

Weishampel, David B., Peter Dodson, and Halszka Osmolska, eds. *The Dinosauria*. Berkeley: University of California Press, 1990. A vast, technical work but the most current and thorough volume on dinosaurs yet written.

Wilford, John Noble. *The Riddle of the Dinosaurs*. New York: Alfred A. Knopf, 1988. An elegantly written and thorough history of dinosaur explorations also touching on the recent work of Horner and extinction theorists.

For a historical or theoretical perspective, and sometimes both, on dinosaurs and evolution, I recommend the many and various books of George Gaylord Simpson (particularly *The Meaning of Evolution*), Edwin H. Colbert (particularly *The Great Dinosaur Hunters and Their Discoveries*), and Stephen Jay Gould (particularly *Bully for Brontosaurus*, *The Panda's Thumb*, and *Hen's Teeth and Horse's Toes*).

Anyone seriously interested in the distant past, what's current in dinosaur research, and what's ahead for all of paleontology ought to belong to the nonprofit Society of Vertebrate Paleontology, c/o W436 Nebraska Hall, University of Nebraska State Museum, Lincoln, NE 68588-0542. Their quarterly *News Bulletin* contains the latest in scientific research worldwide. There are many categories of membership and support.

For lists of recommended books on dinosaurs for children and adults write The Dinosaur Society, the nonprofit organization for dinosaur science, at P.O. Box 171, Newton Lower Falls, MA 02162.

Sources

Wherever possible I have quoted scientists' own descriptions of their research, from my own interviews and those in popular articles as well as scholarly publications.

INTRODUCTION: THE TRUTH ABOUT DINOSAURS

Interviews

Dr. Peter Dodson, University of Pennsylvania; Dr. Stephen Jay Gould, Harvard University; George Olshevsky, San Diego, California.

Publications

Dodson, Peter. "Dinosaurs Defended." *American Scientist* (March-April 1986). Dodson's response to Thomson (see below).

————. "Review. Dinosaur Systematics Symposium." *Journal of Vertebrate Paleontology* 7 (March 1987).

————. "Counting Dinosaurs: How Many Kinds Were There?" *Proceedings, National Academy of Sciences* (1990).

Dodson, Peter, and Susan D. Dawson. "Making the Fossil Record of Dinosaurs." *Modern Geology* (1991).

McFadden, Bruce. "Dinosaurs Revive Geology Departments." *Geo Times* (May 1989).

Morrel, Virginia. "Announcing the Birth of a Heresy." *Discover Magazine* (March 1987).

Thomson, Keith. "Is Paleontology Going Extinct?" *American Scientist* (November-December 1985).

Wing, Scott L., and Hans-Dieter Sues, eds. "Mesozoic and Early Cenozoic." In *The Evolution of Terrestrial Ecosystems,* University of Chicago Press, in press.

CHAPTER 1: APPROACHING THE DINOSAUR FRONTIER

Interviews

Dr. Anna K. Behrensmeyer, Smithsonian Institution; Dr. Philip Currie, Royal Tyrrell Museum; Dr. Peter Dodson, University of Pennsylvania; Dr. John Horner, Museum of the Rockies.

Publications

Begley, Sharon. "How Dinosaurs Lived." *Newsweek* (October 28, 1991).

Behrensmeyer, Anna K. "Taphonomy and the Fossil Record." *American Scientist* 72 (November-December 1984).

————. "Vertebrate Preservation in Fluvial Channels." *Palaeogeography, Palaeoclimatology, Paleoecology* 63 (1988).

Currie, Philip J. "Birdlike Characteristics of the Jaws and Teeth of Troödontid Theropods." *Journal of Vertebrate Paleontology* (March 1987).

Currie, Philip J., and Peter Dodson. "Mass Death of a Herd of Ceratopsian Dinosaurs." In *Third Symposium of Mesozoic Terrestrial Ecosystems, Short Papers,* edited by W.E. Reif and F. Westphal. Tübingen, Germany: Attempton Verlag, 1984.

Dodson, Peter. "Sedimentology and Taphonomy of the Oldman Formation." *Palaeogeography, Palaeoclimatology, Palaeoecology* 10 (1971).

Fiorillo, Anthony R. "Trample Marks: Caution from the Cretaceous." *Current Research in the Pleistocene* 4 (1987).

Mesozoic Terrestrial Ecosystems, fourth symposium. *Occasional Paper of the Tyrrell Museum of Paleontology,* #3 (1987).

"Paleocology of Upper Cretaceous Judith River Formation at Dinosaur Provincial Park." *Occasional Paper of the Tyrrell Museum of Paleontology,* #7 (1988).

Reid, Gordon. *Dinosaur Provincial Park.* Erin, Ontario: The Boston Mills Press, 1986.

Russell, Dale A. *A Vanished World: The Dinosaurs of Western Canada.* Toronto: National Museum of Canada, 1977.

Tyrrell Museum of Paleontology and the Drumheller Valley. Tyrrell Museum, Drumheller, Canada: Wildland Publishing, 1987.

CHAPTER 2: THE SEARCH FOR THE FIRST DINOSAUR

Interviews

Dr. Robert Bakker, Boulder, Colorado; Dr. Jose Bonaparte, Museo de Ciencias Naturales, Argentina; Dr. Jacques Gauthier, California Academy of Sciences; Dr. Stephen Jay Gould, Harvard University; Dr. Paul Sereno, University of Chicago.

Publications

Bakker, Robert T. "Dinosaur Renaissance." *Scientific American* 232 (April 1975).

Bakker, R., and P. Galton. "Dinosaur monophyly and a new class of vertebrates." *Nature* 248 (1974).

Benton, Michael J. "Dinosaurs' Lucky Break." *Natural History* 93 (1984).

———. "More than one event in the late Triassic mass extinction." *Nature* 321 (June 26, 1986).

———. "The Origins of the Dinosaurs." *Modern Geology* 13 (1988). Great Britain: Gordon and Breach Science Publishers.

————. "Origin and Interrelationships of Dinosaurs." In *The Dinosauria,* edited by David Weishampel, Peter Dodson, and Halszka Osmolska. Berkeley: University of California Press, 1990.

Benton, Michael J., and James M. Clark. "Archosaur phylogeny and the relationships of the Crocodylia." In *The Phylogeny and Classification of the Tetrapods, Volume I: Amphibians, Reptiles, Birds,* edited by Michael J. Benton. Oxford, Great Britain: Clarendon Press, 1988.

Bonaparte, Jose F. "Dinosaurs: A Jurassic Assemblage from Patagonia." *Science* 205 (September 28, 1979). One of the most easily accessed of Bonaparte's writings. What follows are a selection from Bonaparte's oeuvre for those with ample language and library resources.

————. "A Horned Cretaceous Carnosaur from Patagonia." *National Geographic Review* (Winter 1985).

————. "Les Dinosaures du Jurassique Moyen de Cerro Condor." *Annales de Paleontologie* (Paris) 72 (1986).

Bonaparte, Jose F., and Fernando Novas. *"Abelisaurus comahuensis,* Carnosauria del Cretacico Tardio de Patagonia." *Ameghiniana* (Buenos Aires) 21, nos. 2–4 (1985).

Bonaparte, Jose F., and Martin Vince. "El Hallazgo del Primer Nido de Dinosaurios Triasicos, Triasico Superior de Patagonia Argentina." *Ameghiniana* (Buenos Aires) 16 (July 1979).

"A Field Trip Bonanza," *The New York Times* (April 9, 1991).

Gasparini, Z., E. Olivero, R. Scasso, and C. Rinaldi. "Un Ankylosaurio Campaniano en el Continente Antartico." *Anais do X Congresso Brasiliero de Paleontologia,* Rio de Janeiro, July 19–25, 1987.

Gauthier, Jacques. "Saurischian Monophyly and the Origin of Birds." In *The Origin of Birds and the Evolution of Flight,* edited by Kevin Padian. *Memoirs of the California Academy of Sciences,* no. 8, San Francisco: California Academy of Sciences, 1986.

Gould, Stephen J. "The Telltale Wishbone." In his *The Panda's Thumb.* New York: W. W. Norton, 1980.

————. "What, If Anything, Is a Zebra." In his *Hen's Teeth and Horse's Toes.* New York: W. W. Norton, 1983.

Halstead, Beverly. "Museum of Errors" correspondence. *Nature* 288 (November 20, 1980). (Replies, including Benton's, and Halstead's further comments, in *Nature* 289, January 1/8, 1981.)

Kluger, Jeffrey. "Tripping over History." *Discover* (December 1990).

Monastersky, Richard. "Chinese bird fossil: Mix of old and new." *Science News* 138 (October 20, 1990).

Padian, Kevin, ed. *The Beginning of the Age of Dinosaurs: Faunal Change Across the Triassic-Jurassic Boundary.* New York: Cambridge University Press, 1986.

Sereno, Paul C. "Basal Archosaurs: Phylogenetic Relationships and Functional Implications." Society of Vertebrate Paleontology Memoir no. 2, *Journal of Vertebrate Paleontology* 11, supp. to no. 4 (December 31, 1991).

―――. "Phylogeny of the Bird-Hipped Dinosaurs." *National Geographic Research* 2, no. 2 (1986).

CHAPTER 3: THE CASE OF THE FLYING DINOSAUR

Interviews

Dr. Robert Bakker, Boulder, Colorado; Dr. Sankar Chatterjee, Texas Tech University; Dr. Peter Dodson, University of Pennsylvania; Dr. Jacques Gauthier, California Academy of Sciences; Dr. Lawrence Martin, University of Kansas; Dr. Kevin Padian, University of California/ Berkeley; Dr. Michael Parrish, University of Kansas; Dr. Paul Sereno, University of Chicago.

Publications

Bock, W. J. "On Extended Wings." *The Sciences* 23 (1983).

Caple, G. R., R. T. Balda, and W. R. Willis. "The Physics of Leaping Animals and the Evolution of Preflight." *American Naturalist* 121 (1983).

Chatterjee, Sankar. "A Primitive Parasuchid Reptile from the Upper Triassic Maleri Formation of India." *Palaeontology* 21 (February 1978).

―――. "Indosuchus and Indosaurus, Cretaceous Carnosaurs from India." *Journal of Paleontology* 52, no. 3 (May 1978).

————. "Phylogeny and classification of thecodontian reptiles." *Nature* 295 (1982).

————. "A New Ornithischian Dinosaur from the Triassic of North America." *Naturwissenschaften* 71 (1984). West Germany: Springer Verlag.

————. *"Postosuchus,* a New Thecodontian Reptile from the Triassic of Texas and the Origin of Tyrannosaurs." *Philosophical Transactions of the Royal Society of London* B 309 (1985).

————. "A New Theropod Dinosaur from India with Remarks on the Gondwana-Laurasia Connection in the Late Triassic." In *Sixth Gondwana Symposium* edited by Gary D. McKenzie. Geophysical Monograph 41. Columbus, Ohio: American Geophysical Union, 1987. See also Chatterjee's paper on Dockum, Texas, vertebrates in the Padian-edited *The Beginnings of the Age of Dinosaurs.*

————. "Cranial anatomy and relationships of a new Triassic bird from Texas." *Philosophical Transactions: Biological Sciences* 322, #1265 (June 29, 1991): 277–346. London: The Royal Society.

Chatterjee, Sankar, and Nicholas Hotton III. "The paleoposition of India." *Journal of Southeast Asian Earth Science* 1, no. 3 (1986).

Dodson, Peter. "Conference Report. International *Archaeopteryx* Conference." *Journal of Vertebrate Paleontology* 5, no. 2 (June 1985).

Gauthier, Jacques. *A Cladistic Analysis of the Higher Systematic Categories of the Diapsida.* Ann Arbor, Michigan: University Microfilms, 1984.

Grande, Lance, and Sankar Chatterjee. "New Cretaceous Fish Fossils from Seymour Island, Antarctic Peninsula." *Palaeontology* 30, pt. 4 (1987).

Hecht, Max, John H. Ostrom, Gunther Viohl, and Peter Wellnhofer. *The Beginnings of Birds.* Willibaldsburg, West Germany: Jura Museum, 1985. The proceedings of the 1984 Eichstätt *Archaeopteryx* conference, with many important papers worth reading for any concerned with bird and bird-flight origins.

Kurzanov, S. M. "Avimimids and the Problem of the Origin of Birds." *Transactions of the Joint Soviet-Mongolian Paleontological Expedition* 31 (1987).

Ostrom, John H. "The ancestry of birds." *Nature* 242 (1973).

————. *Archaeopteryx and the origin of flight. Quarterly Review of Biology* 49 (1974).

————. "Bird flight: how did it begin?" *American Scientist* 67, no. 1 (1979).

Padian, Kevin. "A Functional Analysis of Flying and Walking in Pterosaurs." *Paleobiology* 9 (1983).

————. "The Origins and Aerodynamics of Flight in Extinct Vertebrates." In *Palaeontoology* 28, pt. 3 (1985). See also Padian's essay, "The Case of the Bat-Winged Pterosaur: Typonomic Taxonomy and the Influence of Pictorial Representation on Scientific Perception" in *Dinosaurs Past and Present*, Volume II.

Pennycuick, Colin J. *Mechanical Constraints on the Evolution of Flight.* Memoirs of the California Academy of Sciences, no. 8, 1986.

————. "On the Reconstruction of Pterosaurs and Their Manner of Flight, With Notes on Vortex Wakes." *Biology Review* 63 (1988).

Rayner, J. M. V. "A vortex theory of animal flight, Part 2. The forward flight of birds." *Journal of Fluid Mechanics* 91 (1979).

Sereno, Paul C. "*Lesothosaurus,* 'Fabrosaurids,' and the early evolution of Ornithischia." *Journal of Vertebrate Paleontology* 11, no. 2 (June 1991).

Simpson, George Gaylord. *Penguins: Past and Present, Here and There.* New Haven, Connecticut: Yale University Press, 1976.

————. *Concession to the Improbable: An Unconventional Autobiography.* New Haven, Connecticut: Yale University Press, 1978.

Unwin, David M. "New pterosaurs from Brazil." *Nature* 332 (61633) (1988).

Walker, A. D. "New light on the origin of birds and crocodiles." *Nature* 237 (1973).

Wellnhofer, Peter. *Pterosauria.* Handbuch der Palaeoherpetologie, Teil 19. Stuttgart: Gustav Fischer Verlag, 1978.

————. "New Crested Pterosaurs from the Lower Cretaceous of Brazil." *Mitt. Bayerische Staatssammlung für Paläontologie und historische Geologie* 27 (1987).

CHAPTER 4: THE DINOSAUR TAKEOVER

Interviews

Mr. William Amaral, Harvard University; Dr. Robert Bakker, Boulder, Colorado; Dr. Philip Currie, Royal Tyrrell Museum; Dr. James Farlow, University of

Indiana/Purdue; Dr. Paul Olsen, Lamont Biological Laboratories; Dr. Hans-Dieter Sues, Smithsonian Institution; Dr. Kevin Padian, University of California/Berkeley.

Publications

Alexander, R. McNeill. "Estimates of speeds of dinosaurs." *Nature* 261 (1976).

————. *Dynamics of Dinosaurs & Other Extinct Giants.* New York: Columbia University Press, 1989.

Bird, R. T. *Bones for Barnum Brown: Adventures of a Dinosaur Hunter.* Fort Worth: Texas Christian University Press, 1985. With an interesting introduction by Farlow.

Currie, Philip J., and W. A. S. Sarjeant. "Lower Cretaceous Footprints from the Peace River Canyon, British Columbia, Canada." *Palaeogeography, Palaeoclimatology, Palaeobiology* (1979).

Dodson, Peter. "Review, *The Beginning of the Age of Dinosaurs—Faunal Change Across the Triassic-Jurassic Boundary.*" *Journal of Vertebrate Paleontology* 8, no. 2 (June 1988).

Farlow, James O. "Estimates of Dinosaur Speeds from a New Trackway Site in Texas." *Nature* 294 (1981).

————. *A Guide to Lower Cretaceous Dinosaur Footprints and Tracksites of Paluxy River Valley.* Somervell County, Texas: South Central GSA, Baylor University, 1987.

Gillette, David, ed. *First International Symposium on Dinosaur Tracks and Traces.* Albuquerque: New Mexico Museum of Natural History, 1986. Papers by leading footprint researchers, including Farlow.

Gillette, David G., and Martin G. Lockley. *Dinosaur Tracks and Traces.* New York: Cambridge University Press, 1989. A major compendium of papers on tracks and eggs, largely if not entirely a product of the first footprint conference.

Lockley, Martin G. "The paleobiological and paleoenvironmental importance of dinosaur footprints." *Palaios* 1 (1986).

————. "Review, *A Guide to Lower Cretaceous Dinosaur Footprints and Tracksites of Paluxy River Valley.*" *Journal of Vertebrate Paleontology* 8, no. 1 (March 1988).

—————. *Tracking Dinosaurs.* New York: Cambridge University Press, 1991. A major overview of ichnology.

Monastersky, Richard. "Abrupt Extinctions at End of Triassic." *Science News* (September 5, 1987).

Olsen, Paul E. "Comparative Paleolimnology of the Newark Supergroup: A Study of Ecosystem Evolution." Ph.D. diss., Yale University, 1984. These are but a modest selection of Olsen's voluminous writings on the Newark Supergroup and the Triassic-Jurassic boundary.

Olsen, Paul E., and Peter M. Galton. "Triassic-Jurassic tetrapod extinctions: are they real?" *Science* 197 (1977).

Padian, Kevin, and Paul E. Olsen. "The Fossil Trackway *Pteriaichnus:* Not Pterosaurian, but Crocodilian." *Journal of Paleontology* 58, no. 1 (January 1984). See also contributions by Olsen and Sues to the Padian-edited *The Beginning of the Age of Dinosaurs.*

—————. "Footprints of the Komodo Monitor and the Trackways of Fossil Reptiles." *Copeia* 3 (1984).

Thulborn, Anthony. "Preferred gaits of bipedal dinosaurs." *Alcheringa* 8 (1984).

—————. *Dinosaur Tracks.* London: Chapman and Hall, 1990. A thorough treatment by a scientist with an unconventional perspective.

Thulborn, Anthony, and M. Wade. "Dinosaur stampede in the Cretaceous of Queensland." *Lethaia* 12 (1979).

CHAPTER 5: THE DRAGONS OF DASHANPU

Interviews

Dr. Robert Bakker, Boulder, Colorado; Dr. Dong Zhiming, Institute of Paleontology and Paleoanthropology, Beijing, China; Dr. John McIntosh, Wesleyan University; Dr. Xiao Xi Wu, Chongqing Museum, China.

Publications

Dong Zhiming. "A New Genus of Pachcephalosauria from Laiyang, Shantung." *Vertebrata Palasiatica* 15, no. 4 (1978).

Dong Zhiming and Angela Milner. *Dinosaurs from China.* Beijing: China Ocean Press, Beijing and British Museum of Natural History, 1988.

Mateer, Niall J., J. H. Hartman, Zhen S., Li J., Rao C., Chen J., and Reser P. "Stratigraphy, palaeontology and sedimentology of Cretaceous dinosaur-bearing strata of eastern China." Unpublished report to National Geographic Society, 1988.

The Middle Jurassic Dinosaurian Fauna from Dashanpu, Zigong, Sichuan, vol. 1. Chongqing: Sichuan Scientific and Technological Publishing House, 1984. Contributions from Xiao Xi Wu on stegosaurs, He Xinlu and Cai Kaiji on ornithopod dinosaurs, and others.

Olshevsky, George. "More Sauropods from China." *Archosaurian Articulations* 1, no. 8 (February 1989).

Tanimoto, Masahiro. "Tails of Sauropods from the Sichuan Basin, China." *Archosaurian Articulations* 1, no. 2 (August 1988).

"Special Paper on Dinosaurian Remains of Dashanpu, Zigong, Sichuan." *Journal of Chengdu College of Geology,* Supplement 1 (1983). Preliminary descriptions of digs and discoveries by Xiao Xi Wu and others.

CHAPTER 6: HERESIES, HOT-BLOODS, AND HOLES

Interviews

Dr. Robert Bakker, Boulder, Colorado; Mr. Kenneth Carpenter, Denver Museum of Natural History; Dr. Peter Dodson, University of Pennsylvania; Dr. James Farlow, University of Indiana/Purdue; Dr. Stephen Jay Gould, Harvard University; Dr. John Horner, Museum of the Rockies; Dr. Nicholas Hotton III, Smithsonian Institution; Dr. Armand Ricqlès, Paris, France; Ms. Jill Peterson Rife, Cedaredge, Colorado.

Publications

Alexander, R. MacNeill. *Dynamics of Dinosaurs & Other Extinct Giants.* New York: Columbia University Press, 1989.

Bakker, Robert T. "Dinosaur Physiology and the Origin of Mammals." *Evolution* 25 (1971).

————. "Locomotor Energetics of Lizards and Mammals Compared." *Physiologist* 15 (1972).

————. "Dinosaur feeding behavior and the origin of flowering plants." *Nature* 274 (August 17, 1978).

————. "Review of the Late Cretaceous Nodosauroid Dinosauria." *Hunteria* 1, no. 3 (September 14, 1988).

Bakker, Robert T., Michael Williams, and Philip Currie. "*Nanotyrannus,* a new genus of pygmy tyrannosaur from the Latest Cretaceous of Montana." *Hunteria* 1, no. 5 (1988).

Carpenter, Kenneth. "Review, *Dinosaur Heresies.*" *Delaware Valley Paleontological Society Newsletter* 10, no. 6 (1988).

Coe, M. J., D. L. Dilcher, James O. Farlow, D. M. Jarzen, and Dale A. Russell. "Dinosaurs and land plants." In *The Origins of Angiosperms and Their Biological Consequences,* edited by E. M. Friis, W. G. Chaloner, and P. R. Crane. New York: Cambridge University Press, 1987.

Dodson, Peter. "Review, *A Cold Look at the Warm-Blooded Dinosaurs.*" *American Journal of Science* 281 (1981).

————. "Review, *Dinosaur Heresies.*" *American Scientist* (September-October 1987).

Enlow, D. H., and S. O. Brown. "A Comparative Histological Study of Fossil and Recent Bone Tissues, Part II." *Texas Journal of Science* 9 (1957).

Farlow, James O. "Speculations About the Diet and Foraging Behavior of Large Carnivorous Dinosaurs." *The American Midland Naturalist* 95, no. 1 (January 1976).

————. "Dinosaur Energetics and Thermal Biology." In *The Dinosauria,* edited by David Weishampel, Peter Dodson, and Halszka Osmolska. Berkeley: University of California Press, 1990.

Giffin, Emily. "Review, *The Dinosaur Heresies.*" *Palaios* 2, no. 5 (1987).

Hargens, Alan R., Ronald W. Millard, Knut Peterson, and Kjell Johansen. "Gravitational haemodynamics and oedema prevention in the giraffe." *Nature* 329 (September 1987).

Jerison, Harry J. *Evolution of the Brain and Intelligence.* New York: Academic Press, 1973.

McGowan, Christopher. *Dinosaurs, Spitfires & Sea Dragons.* Cambridge: Harvard University Press, 1991.

Monastersky, Richard. "Reining In a Galloping *Triceratops.*" *Science News* (October 20, 1990).

Ostrom, John H. "Osteology of *Deinonychus antirrhopus,* an unusual theropod from the Lower Cretaceous of Montana." *Bulletin of the Peabody Museum of Natural History* 300, 165 S. (1969).

———. "Terrestrial Vertebrates as Indicators of Mesozoic Climates." *North American Paleontological Convention Proceedings,* Chicago (1969).

———. "*Archaeopteryx* and the Origin of Birds." *Biological Journal of the Linnean Society* 8 (1976).

Peterson, Jill. "Histology as a method of determining ontogenetic development in dinosaurs." Unpublished Romer Prize Lecture, Society of Vertebrate Paleontology, 1985.

Reid, Robert E. H. "Bone and dinosaur 'endothermy.' " *Modern Geology* 11 (1987).

Ricqlès, Armand J. de. "Evolution of endothermy, histological evidence." *Evolutionary Theory* 1 (1974).

———. "Cyclical growth in the long bones of a sauropod dinosaur." *Acta Palaeontologica Polonica* 28 (1983).

Robbins, Jim. "The Real Jurassic Park." *Discover* (March 1991).

Spotila, James R., Michael P. O'Connor, Peter Dodson, and Frank V. Paladino. "Hot and Cold Running Dinosaurs: Body Size, Metabolism, and Migration." *Modern Geology* in press (1991).

Stokes, William Lee. *The Cleveland Lloyd Dinosaur Quarry—Window to the Past.* Washington, D.C.: U.S. Government Printing Office, 1985.

Thomas, R. D. K., and E. C. Olson, eds. *A Cold Look at the Warm-Blooded Dinosaurs.* Boulder, Colorado: Westview Press, 1980. *The* document of the warm-blooded debate from the AAAS conference on the same, featuring important summaries by Spotila, Hopson, Hotton, Ricqlès, and Bakker, among others.

Weaver, J. C. "The improbable endotherm: the energetics of the sauropod dinosaur *Brachiosaurus.*" *Paleobiology* 9 (1983).

CHAPTER 7: LAND OF GIANTS

Interviews

Dr. Peter Dodson, University of Pennsylvania; Dr. David Gillette, Salt Lake City, Utah; Dr. Nicholas Hotton III, Smithsonian Institution; Mr. Jim Jensen, Provo, Utah; Mr. James Madsen, Salt Lake City, Utah; Dr. John McIntosh, Wesleyan University; Mr. Cliff Miles, Provo, Utah; Dr. Wade Miller, Brigham Young University; Mr. Kenneth Stadtman, Brigham Young University.

Publications

Berman, David S., and John S. McIntosh. "Skull and relationships of the Upper Jurassic sauropod *Apatosaurus.*" *Bulletin of the Carnegie Museum of Natural History* 8 (1978).

Bonaparte, Jose. "The early radiation and phylogenic relationships of the Jurassic sauropod dinosaurs, based on vertebral anatomy." In *The Beginning of the Age of Dinosaurs,* edited by Kevin Padian. New York: Cambridge University Press, 1986.

Callison, George, and Helen M. Quimby. "Tiny Dinosaurs: Are They Full Grown?" *Journal of Vertebrate Paleontology* 3, no. 4 (March 1984). See also Callison's essay on tiny dinosaurs in volume 1 of *Dinosaurs Past and Present.*

Chure, Dan, ed. *The Morrison Times,* no. 1 (Spring 1991).

"Dinosaur Bones Are Reported in Wyoming." Associated Press (October 27, 1991).

Dodson, Peter. "Paleoecology of the Sauropods." In *The Dinosauria,* edited by David Weishampel, Peter Dodson, and Halszka Osmolska. Berkeley: University of California Press, 1990.

————. "The Lifestyles of the Huge and Famous." *Natural History* (December 1991).

Dodson, Peter, Anna K. Behrensmeyer, Robert T. Bakker, and John S. McIntosh. "Taphonomy and Paleoecology of the Dinosaur Beds of the Jurassic Morrison Formation." *Paleobiology* 6, no. 2 (Spring 1980).

Galton, Peter M., and James A. Jensen. "A New Large Theropod Dinosaur from the Upper Jurassic of Colorado." *BYU Geology Studies* 26, part 2 (July 1979).

Gillette, David D. "*Seismosaurus halli*, Genus ET Species nova, a New Sauropod Dinosaur from the Morrison Formation (Upper Jurassic/Lower Cretaceous) of New Mexico, USA." *Journal of Vertebrate Paleontology* 11, no. 4 (December 31, 1991).

Jensen, James. "Uncompahgre Dinosaur Fauna: A Preliminary Report." *Great Basin Naturalist* 45, no. 4 (October 1985).

————. "Three New Sauropod Dinosaurs from the Upper Jurassic of Colorado." *Great Basin Naturalist* 45, no. 4 (October 1985).

————. "New Brachiosaur Material from the Late Jurassic of Utah and Colorado." *Great Basin Naturalist* 47, no. 4 (October 1987).

————. "A Fourth New Sauropod Dinosaur from the Upper Jurassic of the Colorado Plateau and Sauropod Bipedalism." *Great Basin Naturalist* 48, no. 3 (April 30, 1988).

Jensen, James, and Kevin Padian. "Small Pterosaurs and Dinosaurs from the Uncompaghre Fauna, Late Jurassic, Western Colorado." *Journal of Paleontology* 63, no. 3 (1989).

Monastersky, Richard. "NAS fossil report: Lacking backbone." *Science News* (October 30, 1989).

————. "Dinosaur Digestive Aids." *Science News* (October 20, 1990).

Olshevsky, George. "Late Jurassic North American Brachiosaurids." *Archosaurian Articulations* 1, no. 2 (August 1988).

————. "Sensational Giant Sauropod Pelvis Found at Dry Mesa." *Archosaurian Articulations* 1, no. 4 (October 1988).

Preston, Douglas. "Fossil Wars." *Omni* (October 1988).

Wilford, John Noble. "Two Teams Find Huge Bones in West." *New York Times* (September 6, 1988).

Interviews

Dr. William Clemens, University of California/Berkeley; Dr. Philip Currie, Royal Tyrrell Museum; Dr. Gordon Curry, University of Glasgow; Mark Goodwin, University of California/Berkeley; Dr. Michael Parrish, University of Kansas; Dr. Thomas Rich, University of Victoria, Australia; Dr. Robert Spicer, Oxford University.

Publications

Axelrod, D. I. "An interpretation of Cretaceous and Tertiary biota in polar regions." *Palaeogeography, Palaeoclimatology, Palaeoecology* 45 (1984).

Barron, E. J. "A warm, equable Cretaceous: the nature of the problem." *Earth-Science Reviews* 19 (1983).

Brouwers, Elisabeth M., William A. Clemens, Robert A. Spicer, Thomas A. Ager, L. David Carter, and William V. Sliter. "Dinosaurs on the North Slope, Alaska: High Latitude, Latest Cretaceous Environments." *Science* 237 (September 25, 1987).

Browne, Malcolm W. "Scientists Study Ancient DNA for Glimpses of Past Worlds." *New York Times* (June 25, 1991).

Charig, Alan J., and Angela C. Milner. "*Baryonyx*, a remarkable new theropod dinosaur." *Nature* 324 (November 27, 1986).

"*Claws*": *The Story (so far) of a Great British Dinosaur,* Baryonyx walkeri. London: British Museum of Natural History, 1987.

Clemens, William A., and C. W. Allison. "Late Cretaceous terrestrial vertebrate fauna, North Slope, Alaska" (abstract). *Geological Society of America Abstracts with Program* 17 (1985).

Currie, Philip. "Long-Distance Dinosaurs." *Natural History* (June 1989).

Curry, Gordon B. "Molecular Palaeontology, New Life for Old Molecules." *Trends in Ecology & Evolution* (June 1987).

————. "Amino Acids and Proteins from Fossils." T. W. Broadhead, ed., *Molecular Evolution and the Fossil Record* 1 (1988).

Browne, Malcolm W. "Scientists Study Ancient DNA for Glimpses of Past Worlds." *New York Times* (June 25, 1991).

Davies, K.L. "Duckbill dinosaurs from the North Slope of Alaska." *Journal of Paleontology* 61 (1987).

Douglas, J. G., and G. W. Williams. "Southern polar forests: the early Cretaceous floras of Victoria and their palaeoclimatic significance." *Palaeogeography, Palaeoecology* 39 (1982).

Flannery, Timothy F., and Thomas H. Rich. "Dinosaur digging in Victoria." *Australian Natural History* 20 (1981).

Gee, Henry. "Dinosaurs from the Deep." *The Times* (London) (October 15, 1988).

Hopson, James A. "Relative brain size and behavior in archosaurian reptiles." *Annual Review of Ecology and Systematics* 8 (1977).

Jerison, Harry J. *Evolution of the Brain and Intelligence.* New York: Academic Press, 1973.

Monastersky, Richard. "Dinosaurs in the Dark." *Science News* (March 19, 1988).

―――――. *"Seismosaurus* proteins: Bone of contention." *Science News* (May 4, 1991).

Parrish, J. Michael, Judith Totman Parrish, J. Howard Hutchison, and Robert A. Spicer. "Late Cretaceous vertebrate fossils from the North Slope of Alaska and implications for dinosaur ecology." *Palaois* 2 (1987).

Parrish, Judith Totman, and Robert A. Spicer. "Late Cretaceous terrestrial vegetation: a near-polar temperature curve." *Geology* (January, 1988).

Paul, Gregory S. "Physiological, Migratorial, Climatological, Geophysical, Survival and Evolutionary Implications of Cretaceous Polar Dinosaurs." *Journal of Paleontology* 62, no. 4 (1988).

Rich, Thomas H. V., and Patricia Rich-Vickers. "Polar Dinosaurs and Biotas of the Early Cretaceous of Southeastern Australia." *National Geographic Research* 5, no. 1 (1989).

Russell, Dale A. "The environments of Canadian dinosaurs." *Canadian Geographic Journal* 87 (1983).

Spicer, Robert A., and Judith Totman Parrish. "Paleobotanical evidence for cool north polar climates in middle Cretaceous time." *Geology* 14 (August 1986).

————. "Evolution of Vegetation and Coal-Forming Environments in the Late Cretaceous of the North Slope of Alaska: A Model for Polar Coal Deposition Times of Global Warmth" (special paper). Geological Society of America (in press).

Spicer, Robert A., Jack A. Wolfe, and Douglas J. Nicholas. "Alaskan Cretaceous-Tertiary floras and Arctic origins." *Paleobiology* 3, no. 1 (1987).

Sullivan, Walter. "Near Polar Finds Offer New Look at Dinosaurs." *New York Times* (November 13, 1990).

Vickers-Rich, Patricia, and Thomas R. Rich. "The Dinosaurs of Winter." *Natural History* (April 1991).

Vickers-Rich, Patricia, Thomas R. Rich, Barbara E. Wagstaff, Jennifer McEwen Mason, C. B. Douthitt, R. T. Gregory, and E. A. Felton. "Evidence for Low Temperatures and Biologic Diversity in Cretaceous High Latitudes of Australia." *Science* 242 (December 28, 1988).

Wagstaff, Barbara E., and Jennifer McEwen Mason. "Palynological Dating of Lower Cretaceous Coastal Vertebrate Localities, Victoria, Australia." *National Geographic Research* 5, no. 1 (1989).

Waldman, M. "Fish from the freshwater Lower Cretaceous of Victoria, Australia, with comments on the paleoenvironment." *Special Papers in Palaeontology* 9 (1971).

Winkler, Dale A., Louis L. Jacobs, James Russell Branch, Phillip A. Murry, William R. Downs, and Patrick Trudel. "The Proctor Lake Dinosaur Locality, Lower Cretaceous of Texas." *Hunteria* 2, no. 5 (January 28, 1988).

Wolfe, Jack A. "Arctic Dinosaurs and Terminal Cretaceous Extinctions" (letter). *Science* 239 (January 1, 1988).

Wolfe, Jack A., and G. R. Upchurch, Jr. "Vegetation, climatic and floral changes at the Cretaceous-Tertiary boundary." *Nature* 324 (1986).

————. "North American nonmarine climates and vegetation during the Late Cretaceous." *Palaeogeography, Palaeoclimatology, Palaeoecology* 60 (1987).

Interviews

Dr. Philip Currie, Royal Tyrrell Museum; Dr. Michael Novacek, American Museum of Natural History; Dr. Dale Russell, National Museum of Natural Sciences, Ottawa, Canada; Dr. Artangerel Perle, Mongolian Academy of Sciences, Institute of Vertebrate Paleontology and Paleoanthropology, Beijing, China.

Publications

Andrews, Roy Chapman. *The New Conquest of Central Asia.* New York: American Museum of Natural History, 1932. His greatest work.

————. *Under a Lucky Star.* New York: Viking Press, 1943.

Barsbold, Rinchen. "Carnivorous Dinosaurs from the Cretaceous of Mongolia." *Transactions of the Joint Soviet-Mongolian Paleontological Expedition* 19 (1983). This series provides a voluminous record—unfortunately for most of us, almost entirely in Russian—of Soviet-Mongolian paleontology in the Gobi. English translations of this and other papers are available, for a hefty fee, from Samuel Welles, emeritus professor of paleontology, University of California/Berkeley.

Barsbold, Rinchen, and Artangerel Perle. "Segnosauria, a New Infraorder of Carnivorous Dinosaurs." *Acta Paleontologica Polonica* 25, no. 2 (1980).

"Gobi Diary: A Sedimental Journey." *New York Times Magazine* (November 10, 1991).

Gradzinski, Ryszard, and Tomasz Jerzykiewicz. "Dinosaur and mammal-bearing Aeolian and Associated Deposits of the Upper Cretaceous in the Gobi Desert." *Sedimentary Geology* 12 (1974).

Jerzykiewicz, Tomasz. "Sediments and Dinosaurs in Mongolia." *Dinogramme* 3, no. 1 (1989). Friends of the Tyrrell Museum of Paleontology Foundation.

————. "1988 Sino-Canadian Dinosaur Project Expedition Successful in Inner Mongolia." *Geos* 18, no. 4 (fall 1989).

Jerzykiewicz, Tomasz, and R. A. Sweet. "Semiarid floodplain as a paleoenvironmental setting of the upper Cretaceous dinosaurs: sedimentological evidence from Mongolia and Alberta." *Fourth Symposium on Mesozoic Terrestrial Ecosystems.* Occasional Paper #3 of the Tyrrell Museum of Paleontology (1987).

Kielan-Jaworowska, Zofia. *Hunting for Dinosaurs.* Cambridge: MIT Press, 1969.

Maryanska, Teresa, and Halszka Osmolska. "Pachycephalosauria, a new suborder of ornithischian dinosaurs." In *Results Polish-Mongolian Paleontological Expedition V. Paleontologica Polonica* 30 (1967–71), edited by Zofia Kielan-Jaworowska. This entire publication is highly informative and contains considerable significant dinosaur research by the coauthors.

Perle, Artangerel, Teresa Maryanska, and Halszka Osmolska. *"Goyocephale lattimorei,* a new flat-headed pachycephalosaur from the Upper Cretaceous of Mongolia." *Acta Palaeontologica Polonica* 27, nos. 1–4 (1982).

Tumanova, T. A. "The Armored Dinosaurs of Mongolia." *Transactions of the Joint Soviet-Mongolian Paleontological Expedition* 32 (1987).

Valentine, James W., and Eldridge Moores. *Global Tectonics and the Fossil Record.* Chicago: University of Chicago Press, 1972.

Weishampel, David B., and John R. Horner. "The Hadrosaurid Dinosaurs from the Iren Dabasu Fauna." *Journal of Vertebrate Paleontology* 6, no. 1 (March 1986).

Wilford, John Noble. "After 60 Years, Scientists Return to Fossil Paradise of the Gobi." *New York Times* (July 31, 1990).

CHAPTER 10: EGGS, BABIES, AND NEW DINOSAURS

Interviews

Dr. Peter Dodson, University of Pennsylvania; Dr. Daniel Grigorescu, Bucharest, Romania; Mr. Karl Hirsch, Boulder, Colorado; Dr. John Horner, Museum of the Rockies; Mr. Andrew Leitch, Toronto, Canada; Dr. David Weishampel, Johns Hopkins University.

Publications

Begley, Sharon. "How Dinosaurs Lived." *Newsweek* (October 28, 1991).

Benton, Michael J. "Bringing Up Baby." *Nature* 334 (August 18, 1988).

Coombs, Walter P., Jr. "The Status of the Dinosaurian Genus *Diclonius* and the Taxonomic Utility of Hadrosaurian Teeth." *Journal of Paleontology* 62, no. 5 (1988).

Dodson, Peter. "Taxonomic Implications of Relative Growth in Lambeosaurine Hadrosaurs." *Systematic Zoology* 24 (1975).

————. "*Avaceratops lammersi:* A New Ceratopsid from the Judith River Formation of Montana." *Proceedings of the Academy of Natural Science of Philadelphia* 138, no. 2 (1986).

Fiorillo, Anthony R. "Significance of Juvenile Dinosaurs from Careless Creek Quarry, Wheatland County, Montana." Fourth Symposium on Mesozoic Terrestrial Ecosystems. Occasional Paper #3 of the Royal Tyrrell Museum, 1987.

————. "The vertebrate fauna from the Judith River Formation of Wheatland and Golden Valley counties, Montana." *Mososaur, the Journal of the Delaware Valley Paleontological Society* (1989).

Hirsch, Karl F., and Mary J. Packard. "Review of Fossil Eggs and Their Shell Structure." *Scanning Microscopy* 1, no. 1 (1987).

Hopson, James A. "The evolution of cranial display structures in hadrosaurian dinosaurs." *Paleobiology* 1, no. 1 (Winter 1975).

Horner, John R. "A New Hadrosaur, From the Upper Cretaceous Judith River Formation of Montana." *Journal of Vertebrate Paleontology* 8, no. 3 (September 1988). Horner keeps finding, and naming, more dinosaurs. More to come.

Horner, John R., and David B. Weishampel. "A comparative embryological study of two ornithischian dinosaurs." *Nature* (March 17, 1988).

Horner, John R., and James Gorman. *Digging for Dinosaurs.* New York: Workman Publishers, 1988. This is a compact and highly entertaining review of Horner's research. Also well worth reading in his essay on ecological and behavior implications of a dinosaur nest site in volume 2 of *Dinosaurs Past and Present,* summarizing finds he's presented in many publications.

Kurzanov, S. M., and K. E. Mikhailov. "The finding of dinosaur eggshells in the Lower Cretaceous of Mongolia." In *Dinosaur Tracks and Traces,* edited by David D. Gillette and Martin G. Lockley. New York: Cambridge University

Press, 1989. One of four papers on dinosaur eggshell in this volume, along with works by Hirsch, Jain, and Mateer.

Mikhailov, Konstatin. "The principal structure of the avian eggshell: data of SEM studies." *Acta Zoologica* (Cracow) 30, no. 5 (1987). One of the author's few works in English.

Monastersky, Richard. "Boom in Cute Baby Dinosaur Discoveries." *Science News* (October 22, 1988).

Ostrom, John H. "Were Some Dinosaurs Gregarious?" *Palaeogeography, Palaeoclimatology, Palaeoecology* 11 (1972).

Powell, Jaime Eduardo. "Hallazgo de Huevos Asignables a Dinosaurios Titanosauridos de la Provincia de Río Negro, Argentina." *Il Jornadas Argentinas de Paleontologica de Vertebrados Resumenes* (San Miguel de Tucumán) 16–18 (May 1985).

Sahni, A., R. S. Rana, and G. V. R. Prasad. "SEM studies of thin eggshell fragments from the Intertrappeans of Nagpur and Asifabad, Peninsular India." *Journal of the Paleontological Society of India* 29 (1984).

Shipman, Pat. "Dinosaur Nests: Bringing Up Baby." *Discover* (August 1988).

Srivastava, S., D. M. Mohabey, Ashok Sahni, and S. C. Pant. "Upper Cretaceous Dinosaur Egg Clutches from the Kheda District." *Sonder-Abdruck aus Palaeontographica* (Stuttgart) 1986).

Vianey-Liaud, Monique, Sohan L. Jain, and Ashok Sahni. "Dinosaur Eggshells from the Late Cretaceous Intertrappean and Lameta Formations." *Journal of Vertebrate Paleontology* 7, no. 4 (December 1987).

Weishampel, David B. "The Nasal Cavity of Lambeosaurine Hadrosaurids: Comparative Anatomy and Homologies." *Journal of Paleontology* 55, no. 5 (1981).

————. "Evolution of Jaw Mechanisms in Ornithopod Dinosaurs." *Advances in Anatomy, Embryology, and Cell Biology* 87 (1984). Berlin: Springer-Verlag.

Weishampel, David B., and Reif Wolf-Ernst. "The Work of Franz Baron Nopcsa: Dinosaurs, Evolution, and Theoretical Tectonics." *Jahrbuch der Geologischen Bundesanstalt* 127 (Vienna) (August 1984) 187–203.

Weishampel, David B., Daniel Grigorescu, and David B. Norman. "The Dinosaurs of Transylvania." *National Geographic Research & Exploration* 7, no. 2 (1991).

Wilford, John Noble. "Dinosaur Eggs Yield Glimpse of Their Embryos." *New York Times* (May 3, 1988).

CHAPTER 11: DINOSAURS AT THE BRINK—AND BEYOND?

Interviews

Dr. Robert Bakker, Boulder, Colorado; Dr. William Clemens, University of California/Berkeley; Dr. Peter Dodson, University of Pennsylvania; Dr. Leo Hickey, Yale University; Dr. Kirk Johnson, Denver Museum of Natural History; Dr. Paul Olsen, Lamont Doherty Laboratories; Dr. Kevin Padian, University of California; Dr. Keith Rigby, Jr., Notre Dame University; Dr. Jack Wolfe, United States Geological Survey, Denver, Colorado.

Publications

Alvarez, Luis, Walter Alvarez, Frank Asaro, and Helen Michel. "Extraterrestrial cause for the Cretaceous-Tertiary extinction." *Science* 208 (1980).

Alvarez, Walter, Erle G. Kauffman, F. Surlyk, L. W. Alvarez, Frank Asaro, and Helen Michel. "Impact theory of mass extinctions and the invertebrate fossil record." *Science* 223 (1984).

Alvarez, Walter, and Frank Asaro. "An Extraterrestrial Impact." *Scientific American* (October 1990).

Archibald, J. D. "Stepwise and non-catastrophic extinctions in the Western Interior of North America: testing observations in the context of a historical science." *Memoirs of the Society of Geology of France* 150 (1987).

Archibald, J. D., R. F. Butler, E. H. Lindsay, William A. Clemens, and Lowell Dingus. "Upper Cretaceous-Paleocene biostratigraphy and magnetostratigraphy, Hell Creek and Tullock Formations, northeastern Montana." *Geology* 10 (1982).

Argast, S., James Farlow, R. M. Gabet, and D. L. Brinkman. "Transport-induced abrasion of fossil reptilian teeth: implications for the existence of Tertiary dinosaurs in the Hell Creek Formation, Montana." *Geology* 15 (1987).

Bakker, Robert T. *Dinosaur Heresies.* New York: William Morrow, 1986.

Basu, Asish, Sankar Chatterjee, and Dhirah Rudra. "Shock Metamorphism in Quartz Grains at the Base of the Deccan Traps: Evidence for Impact-Triggered Flood Basalt Volcanism at the Cretaceous-Tertiary Boundary." *Abstract, 1988 Fall Meeting, American Geophysical Union,* San Francisco, 1988.

Bates, Robin. Script for program #4, *Dinosaurs,* WHYY Television, Philadelphia, to air 1992.

Bohor, Bruce F., E. E. Ford, P. J. Modreski, and D. M. Triplehorn. "Mineralogical evidence for an impact event at the Cretaceous-Tertiary dinosaur extinction." *Science* 234 (1984).

Buffetaut, Eric, Jean-Jacques Jaeger, and Jean-Michel Mazin. "Extinctions in Vertebrate History." *Mémoires de la Société Géologique de France* 150 (1987).

Byars, Carlos. "Scientists propose differing meteor impact sites." *Houston Chronicle* (March 16, 1990).

Carpenter, Kenneth, and Breithaupt, Brent. "Latest Cretaceous Occurrence of Nodosaurid Ankylosaurs in Western North America and the Gradual Extinction of the Dinosaurs." *Journal of Vertebrate Paleontology* 6, no. 3 (September 1986).

Chandler, David L. "What killed dinosaurs? Both sides may be right." *Boston Globe* (May 2, 1988).

————. "Caribbean Crater Supports Theory of Dinosaur Deaths." *Boston Globe* (May 18, 1990).

————. "Fossils point to day summer turned to winter, 66 million B.C." *Boston Globe* (August 1, 1991).

————. "Last Chapter of Dinosaur Mystery?" *Boston Globe* (October 28, 1991).

Chatterjee, Sankar. "A possible K-T impact site at the India-Seychelles boundary." *Abstract, Lunar and Planetary Science Conference, XXI,* 1990.

Clemens, Elisabeth S. "Of Asteroids and Dinosaurs: The Role of the Press in the Shaping of Scientific Debate." *Social Studies of Science* (London) 16 (1986).

Clemens, William A. "Evolution of the Terrestrial Vertebrate Fauna During the Cretaceous-Tertiary Transition." In *Dynamics of Extinction,* edited by D. K. Elliott. New York: John Wiley & Sons, Inc., 1986.

Courtillot, Vincent E. "A Volcanic Eruption." *Scientific American* (October 1990).

Dingus, Lowell. "Effects of stratigraphic completeness on interpretations of extinction rates across the Cretaceous-Tertiary boundary." *Journal of Paleontology* 58 (1984).

Dodson, Peter, and Leonid P. Tatarinov. "Dinosaur Extinction." In *The Dinosauria,* edited by David Weishampel, Peter Dodson, and Halszka Osmolska. Berkeley: University of California Press, 1990.

Dricks, Victor. "Dinosaurs 'Broiled' UOFA Scientists Say." *The Phoenix Gazette* (January 20, 1990).

Erben, H. K., J. Hoefs, and K. H. Wedephol. "Paleobiological and isotopic studies of eggshells from a declining dinosaur species." *Paleobiology* 5, no. 4 (1979).

Farlow, James O., and D. L. Brinkman. "Serration Coarseness and Patterns of Wear of Theropod Dinosaur Teeth." *Abstract, 21st Annual Meeting, South-Central Section, Geological Society of America,* Waco, Texas, 1987.

Fassett, James E. "Dinosaurs in the San Juan Basin, New Mexico, may have survived the event that resulted in creation of an iridium-enriched zone near the Creatceous/Tertiary boundary." *Geological Society of America Special Paper 190* (1982).

Fassett, James E., Spencer G. Lucas, and Michael F. O'Neill. "Dinosaurs, pollen and spores and the age of the Ojo Alamo Sandstone, San Juan Basin, New Mexico." *Geological Society of America Special Paper 209* (1987).

Gee, Henry, and Rory Howlett. "Helping with Inquiries." *Nature* 336 (December 15, 1988).

Giffin, Emily B., Diane L. Gabriel, and Rolf E. Johnson. "A New Pachycephalosaurid Skull from the Cretaceous Hell Creek Formation of Montana." *Journal of Vertebrate Paleontology* 7, no. 4 (December 1987).

"Global Fire Is Linked to Dinosaur's Demise." *New York Times* (August 30, 1988).

Gore, Rick. "Extinctions." *National Geographic* (June 1989).

Hallam, Anthony. "End-Cretaceous Mass Extinction Event: Argument for Terrestrial Causation." *Nature* (November 27, 1987). The extinction literature is vast and fast moving, but this was, when written, a thorough survey of current findings, with a noncatastrophist conclusion.

Haq, B. U., J. Hardenbol, and P. R. Vail. "Chronology of fluctuating sea levels since the Triassic." *Science* 235 (1987).

Hecht, Jeff. "Evolving theories for old extinctions." *New Scientist* (November 1988).

Hodge, Carle. "Dinosaurs Died in Fireball Rain, Scientists Think." *Arizona Republic* (January 19, 1990).

Hsu, Kenneth J. *The Great Dying: Cosmic Catastrophe, Dinosaurs and the Theory of Evolution.* Orlando, Florida: Harcourt Brace Jovanovich, 1986.

Hut, Piet, Walter Alvarez, W. P. Elder, T. Hansen, Erle G. Kaufman, G. Keller, E. M. Shoemaker, and P. R. Weissman. "Comet showers as a cause of mass extinctions." *Nature* 329 (1987).

Jaroff, Leon. "At Last, the Smoking Gun." *Time* (July 1, 1991).

Kerr, Richard A. "Snowbird II: Clues to Earth's Impact History." *Science* 242 (December 9, 1988).

Leary, Warren E. "Big Asteroid Passes Near Earth Unseen in a Rare Close Call." *New York Times* (April 20, 1989).

Mathur, U. B., and S. C. Pant. "Sauropod Dinosaur Humeri from Lameta Group (Upper Cretaceous-?Paleocene) of Kheda District, Gujarat." *Journal of the Paleontological Society of India* (Lucknow) 29 (December 1986).

McKenna, Malcolm C. "Collecting microvertebrate fossils by washing and screening." In *Handbook of Paleontological Techniques,* edited by Bernie Kummel and David Raup. San Francisco: W. H. Freeman and Co., 1965.

Orth, Charles J., James S. Gilmore, Jere D. Knight, Charles L. Pillmore, Robert H. Tschudy, and James E. Fassett. "An Iridium Abundance Anomaly at the Palynological Cretaceous-Tertiary Boundary in Northern New Mexico." *Science* 214 (1981).

Padian, Kevin, and William A. Clemens. "Terrestrial Vertebrate Diversity: Episodes and Insights." In *Phanerozoic Diversity Patterns: Profiles in Macroevolution,* edited by James W. Valentine. Princeton, New Jersey: Princeton University Press, 1985.

Rampino, Michael R., and R. B. Strothers. "Flood basalt volcanism during the past 250 million years." *Science* 241 (1988).

Raup, David M. "Mass Extinction: A Commentary." *Paleontology* 30, pt. 1 (1987).

Raup, David M., and George E. Boyajian. "Patterns of generic extinction in the fossil record." *Paleobiology* 14, no. 2 (1988).

Rigby, J. Keith, Jr. "Death by Degrees." *Earthwatch* 8, no. 6 (June 1989).

Rigby, J. Keith, Jr., Karl R. Newman, Jan Smith, and Robert E. Sloan. "Dinosaurs from the Paleocene Part of the Hell Creek Formation, McCone County, Montana." *Palaios* 2 (1987).

Ritter, Malcolm. "Glassy Blobs Add to Evidence That Asteroid Killed Off Dinosaurs." Associated Press (February 7, 1991).

Rogers, Michael. "The Death of the Dinosaurs." *Newsweek* (December 19, 1988).

Russell, Dale A. "The Gradual Decline of the Dinosaurs—Fact or Fallacy?" *Nature* 307 (January 26, 1984).

———. "A note on the terminal Cretaceous nodosaurids of North America." *Journal of Vertebrate Paleontology* 7 (1987).

Saito, T., T. Yamanoi, and T. Kaiho. "End-Cretaceous devastation of terrestrial flora in the boreal Far East." *Nature* 323 (1986).

Sheehan, Peter, and Daniel Fastovsky. "Sudden Extinctions of the Dinosaurs in the Upper Cretaceous Great Plains, USA." *Science* (in press, 1991).

Sloan, Bob, Keith Rigby, L. Van Valen, and Diane Gabriel. "Gradual extinction of dinosaurs and the simultaneous radiation of ungulate mammals in the Hell Creek Formation of McCone County, Montana." *Science* 232 (1986).

Smit, J., and S. Van der Kaars. "Terminal Cretaceous extinctions in the Hell Creek area, Montana: Compatible with catastrophic extinctions." *Science* 223 (1984).

Spicer, Robert A., R. J. Burnham, P. Grant, and H. Glicke. "*Pityrogramma calomelanos,* the primary post-eruption colonizer of Volcan Chichonal, Chiapas, Mexico." *American Fern Journal* 75 (1985).

Stanley, Steven. *Extinction.* New York: Scientific American Library, 1987.

Sullivan, Robert. "Dino Wars." *Dartmouth Alumni Magazine* (February 1990).

Teichert, Curt. "Extinctions and Extinctions." *Palaios* 2, no. 5 (1987).

Thomson, Keith Stewart. "Anatomy of the Extinction Debate." *American Scientist* 76 (January–February 1988).

Wilford, John Noble. "For Dinosaur Extinction Theory, a 'Smoking Gun.' " *New York Times* (February 7, 1991).

Wolbach, Wendy S., I. Gilmour, Edward Anders, C. J. Orth, and R. R. Brooks. "Global fire at the Cretaceous-Tertiary boundary." *Nature* 334 (1988).

Wolfe, Jack. "Paleobotanical Evidence for a June 'Impact Winter' at the K-T boundary." *Nature* (August 1, 1991).

Wolfe, Jack A., and Gary R. Upchurch. "Vegetation, climatic and floral changes at the Cretaceous-Tertiary boundary." *Nature* 324 (1986).

————. "North American nonmarine climate and vegetation during the Late Cretaceous." *Palaeogeography, Palaeoclimatology, Palaeoecology* 61 (1987).

INDEX

Academia Sinica, 134
acid rain, 295, 296
Aepyornis, 269
Africa, 118, 140, 143, 147, 188, 206,
    269, 273, 300
  Cretaceous period in, 214
Alaska, 16, 20, 201, 215–21, 277, 298,
    302
  Cretaceous climate of, 215–16
  gradualism theory and, 218–20
  lightweight fossils from, 223
  migration evidence of, 219
  Pachyrhinosaurus found in, 218,
    221
  plant fossil record of, 216
Alberta, Canada, 27–28, 240
Albertosaurus, 39, 41, 48
All About Dinosaurs (Andrews), 203
Allosaurus, 186, 187, 196, 201, 206
"Alternative to Dinosaur Endothermy,
    An: The Happy Wanderers"
    (Hotton), 220
Alvarez, Luis, 44, 119, 203, 219, 288,
    290, 293, 294, 301
Alvarez, Walter, 44, 116, 290, 293,
    300
Amaral, William, 111–12

American Geophysical Union, 295,
    299–300
American Museum of Natural History,
    39, 62–63, 89, 192, 193, 231,
    253, 254, 255, 264
  Andrews and, 228–29
  Central Asiatic Expedition of, 226
American Scientist, 154
Ames Research Center, 293
amino acids, 222–23
amniotes, 81
Anders, Ed, 295
Andrews, Roy Chapman, 203, 238,
    242, 245, 247, 251, 254, 255,
    267
  dinosaur embryos reported by,
    269–70
  Flaming Cliffs found by, 230–31
  fossil egg sold by, 231
  Gobi expedition of, 226, 228–31
  Iren Dabasu quarry visited by, 228,
    229–30
  Kielan-Jaworowska's refutation of,
    234
  reputation of, 228
ankylosaurs, 245–46, 304
Antarctica, 16, 17, 89, 210, 212

Apatosaurus, 19, 38, 81, 142, 151, 185, 192, 194
  McIntosh's identification of, 193
Archaeopteryx, 80, 84, 85, 86, 96, 100, 152
  privately held specimen of, 195
  Protoavis and, 87, 93, 95
Archaeopteryx lithographica, 82
archosaurs, 66
  defined, 24
  "headless" femur of, 67
  peg-and-socket ankle bones of, 66–68, 76
Argentina, 46, 54, 140, 267, 269
  Ischigualasto Formation of, 51, 57–62, 70–73, 76, 136, 179
  range of fossils in, 59
Armstrong, Harley, 194
arsenic, 145–46
Asaro, Frank, 290
Asia, 17, 118, 143, 190, 218, 226, 229–30, 246, 273, 277
  Cretaceous period in, 214
  first dinosaur discovery in, 228
  lack of horned dinosaurs in, 253–54
Asian-American migration, 228–30, 234–35
  in Cretaceous period, 252–54
  Kielan-Jaworowska's doubt of, 235
asteroids, 293, 295
astronomical cycles, 114–15
Attenborough, David, 309n
Aulenback, Kevin, 242, 247, 252
Australia, 16, 17, 38, 125, 139, 199–215, 219, 302
  ancient polar climate of, 210–12
  Dinosaur Cove site in, 201–7, 210, 213
  dinosaur fauna of, 200–201
  flowering plant evolution in, 214–15
  iguanodon of, 213
  Koonwarra site in, 210
  "living fossil" in, 212
  Slippery Rock site in, 204
Avimimus, 253

Baird, Don, 110
Bakker, Robert, 43, 45, 49, 79, 84, 85, 125, 142, 145, 147, 188, 189, 261, 288, 299, 309n

background of, 151
career of, 150, 151–52
dinosaur behavior as proposed by, 161–62
dinosaur-bird debate and, 84, 85
Dodson's criticism of, 153–56, 162
Dong compared with, 133
feeding strategy as proposed by, 160–61
Gould's assessment of, 152
growth rates studied by, 157–58
lecturing style of, 150–51
McIntosh as recalled by, 192
maverick stance of, 154
Ostrom and, 152
Peterson's work praised by, 164
professional criticism of, 154–56
speed adaptation strategy proposed by, 159–60
speed calculated from footprint by, 160
Stahmer's assessment of, 149
Baluchitherium, 230
"Bang Gang, The" (Nipp), 287
Baryonyx, 214
Basu, Asish, 299–300
Bayan Manduhu site, 242–48
  Dinosaur Project at, 245–48
  geological formation of, 248–50
  importance of, 247–48
Beard, Henry, 23
Behrensmeyer, Kay, 188, 189
Belgium, 38, 213–14
Benton, Michael, 105, 106, 117
Berman, David, 192
Beuche, Antoinette, 204
biomarkers, 222–23
bird-hipped dinosaurs (ornithischians):
  cladistics and, 69
  defined, 25
  eggs of, 268
  evolution of, 67–68
  Herrerasaurus and, 72
  proliferation of, 103–4
birds, 20, 301
  ankle bones of, 66–68
  origin and development of, 81–83
  Troödon's similarity to, 36, 84
  see also dinosaur-bird debate
blue-green algae, 65

Bonaparte, Jose Napoleon, 54, 58, 59, 69, 70, 73, 191, 269, 309*n*
Bourgeois, Joanne, 299
brachiosaurids, 193–94
Brachiosaurus, 141, 172, 176, 184, 187
brain size, intelligence and, 210
Brandvold, Marion, 263
Brazil, 17, 59, 62
Brigham Young University (BYU), 136, 169, 174, 183, 184
  fossils sold by, 197
  Jensen's finds as stored at, 185–86
British Museum, 63, 69, 195, 282, 309
Britt-Surman, Michael, 29
"Brontosaurus," *see Apatosaurus*
Brown, Barnum, 39, 89, 264
Buckland, William, 37, 144
Buenos Aires Museum of Sciences, 59
Buffetaut, Eric, 146–47
Bureau of Land Management, U.S., 180, 196
BYU, *see* Brigham Young University

California Academy of Science, 69
Camarasaurus, 193, 194
Camp, Charles, 131
"Camp"-osaur Quarry, 264
Canada, 17, 27–28, 106, 187, 268, 302
  Chinese paleontologists in, 239–40
Canadian Dinosaur Rush, 39
"Cape Paterson Claw," 201
Carboniferous era, 65
Carnegie Museum, 192, 193, 194
Carpenter, Kenneth, 154
Carrano, Matthew, 276
catastrophic impact theory, 44–45, 106–7, 117, 200
  asteroids and, 293
  comets and, 294
  crater as evidence for, 291–93
  deciduousness and, 296–97
  Dong's view of, 289
  Gould on, 288
  iridium and, 119, 290, 298, 299
  Manicouagan site and, 118–19
  Olsen's support for, 117–18, 120–21
  periodicity and, 294
  polar adaptations and, 219
  public acceptance of, 291
  scientific acceptance of, 305
  tektites and, 292

Triassic-Jurassic boundary and, 119
Wassons Bluff and, 118
  see also gradual extinction theory; mass extinction theory; step-wise extinction theory
Central Asiatic Expedition, 226
"Centrosaurus" *(Eucentrosaurus),* 46–47
ceratopsians, *see* horned dinosaurs
Chasmosaurus, 65
Chatterjee, Sankar, 87–101, 135
  background of, 89–90
  on Basu's volcanism theory, 300
  early career of, 91–93
  fossil-finding knack of, 90
  Hotton's view of, 90, 99
  ostrich-mimic fossil of, 100–101
  Padian's criticism of, 99
  Protoavis analysis by, 94–97
  Protoavis discovered by, 87–89, 93–94
  Protoavis treatise of, 98–99
  at SVP meetings, 100
chemical dating, 116–17
Chengdu College of Geology, 145
Chicxulub crater, 292–93, 294
China, 16, 17, 38, 46, 59, 63, 84, 89, 124, 132, 203, 231
  dinosaur richness of, 104–8
  dinosaurs in culture of, 128–29
  first dinosaur bones found in, 129–31
  fossil policy of, 138
  paleontology in, 134–36
  yellow loess of, 227
Chinese-Canadian Dinosaur Project, 134
  baby dinosaurs found by, 245–47
  at Bayan Manduhu, 245–48
  Currie with, 225–26, 228, 238–40, 242, 245–46, 247, 252, 253
  Dong with, 225–28, 240, 242
  eggs found by, 267
  fossil eggshells found by, 247
  at Iren Dabasu site, 251–55
  Russell and, 240, 242
  tunneling burrows found by, 247
Chu Hsi, 129
cladistics, 65, 73, 76, 191, 235
  development of, 68
  dinosaur-bird debate and, 84, 85

cladistics *(cont.)*
  lizard-hipped dinosaurs and, 69
  role of time and, 85
Clemens, Elizabeth, 291
Clemens, William A., 216–20, 291,
    295, 301–2, 304
climate:
  of ancient Australia, 210–12
  ancient lakes and, 115
  of ancient Nova Scotia, 118
  in Cretaceous Alaska, 215–16
  in Jurassic period, 186
  mass extinction and, 115, 289
  of Pangaea, 53–54
  in Permian period, 66
  polar, 210
  rapid growth and, 165
  of Wassons Bluff, 118
coevolution, 153
Colbert, Edwin, 62
*Cold Look at the Warm-Blooded
    Dinosaurs, A* (Dodson), 220
Colgate University, 231
Colorado, 17, 187, 221, 283
Colorado, University of, 150, 154
Colville River, Alaska, 216, 221
comets, 294
commercial collecting, 195–97
Como Bluff site, 153
"competition" theory, 104–5
Compsognathus, 81
convergent evolution, 68, 82–83, 121
  dinosaur-bird debate and, 84, 85
Cope, Edwin Drinker, 37–38, 136, 276
"cow-turtles," 54
Cretaceous period, 63, 66, 84, 95, 118,
    140, 186, 187, 190, 201, 205,
    289, 305
  in Africa, 214
  in Asia, 214
  Asian-American migration in,
    252–54
  Australian dinosaurs of, 200, 206
  climate of, 215–16, 296
  defined, 24
  dinosaur evolution in, 212, 214
  diversity after, 296–97
  flourishing of duckbills in, 273
  flowering plants in, 214–15
  fossil record of, 212–13
  Horner's concept of habitat of, 258

inland sea of, 279, 282–84
  proliferation of dinosaurs in, 104
Cretaceous-Tertiary (K-T) Boundary,
    119, 292, 297–98, 300
  dinosaur diversity and, 302–4
  iridium at, 119, 298, 299
  sedimentation at, 301
Crichton, Michael, 222, 223, 309*n*
Cronkite, Walter, 218
Cultural Revolution, 134–36
Currie, Phil, 42, 48, 49, 222, 251, 254,
    255, 278, 279, 309*n*
  with Dinosaur Project, 225–26, 228,
    238–40, 242, 245–46, 247, 252,
    253
  eggs and nests found by, 46,
    267–68, 271–72
  herding studied by, 46–47
  migration favored by, 221
  Troödon and, 29–36
Curry, Gordon, 222, 223
Cuvier, Georges, 36–37

Darwin, Charles, 81, 98
Dashanpu quarry, 124, 141–42, 187,
    190
  arsenic concentrations in, 145–46
  dating of, 146–47
  described, 128
  Dong and Xiao at, 138
  geology of, 136–38
  quality of fossils in, 124–28
Datousaurus, 128
"Death Star," 44
Deccan Traps, 289, 295, 300
deciduousness, 296–97
Deinocheirus, 234
Deinonychus, 42–43, 96, 152
  dinosaur-bird debate and, 83
  Ostrom's description of, 157
  Ostrom's finding of, 214
"dental battery," 273
Denver Museum of Natural History,
    154, 297
Digging Dinosaurs (Horner and
    Gorman), 162, 165
Dimetrodon, 65
  plastic replica of, 309
Dinomania, prevalence of, 309
dinosaur-bird debate, 20
  Bakker and, 84, 85

cladistics and, 84, 85
convergent evolution and, 84, 85
Deinonychus and, 83
Gauthier and, 85, 86
Martin and, 85–86
Ostrom and, 83–84, 152
Protoavis and, 86–89, 93–100
Sereno and, 84
Troödon and, 36, 84
Dinosaur Cove, 201–7, 210, 213
ancient fauna of, 206
Dinosaur Heresies (Bakker), 153–54
Dinosauria, 37, 84
dinosaur paleontology:
Cope-Marsh rivalry and, 37–38
evolutionary origins and, 57
new finds and, 16–17
other disciplines and, 106
sedimentology and, 248–51
slow pace of, 70
technological advances and, 17–18,
39, 57
underfunding of, 307–9
zoological nomenclature and,
172–74
Dinosaur Project, *see*
Chinese-Canadian Dinosaur
Project
Dinosaur Provincial Park, 39, 221, 274
Currie's studies in, 46
geology of, 40
fossil yield of, 28
Troödon discovered in, 34
"Dinosaur Renaissance" (Bakker), 152
dinosaurs:
ankle structure of, 66–68
astronomical cycles and, 114–15
blood pressure of, 141
Chinese terms for, 128
classification of, 65
common perception of, 18–20, 150
common rootstock of, 57
definition of, 67–68
disparity of nestling sizes of, 163
eggs of, *see* eggs, dinosaur
feeding strategy of, 160–61
first appearance of, 81
first scientific discoveries of, 36–37
first use of term, 37
foraging rates for, 159–60
growth of, *see* growth of dinosaurs

hip structure of, 67–68
known genera of, 16
lifespan of, 271
mania for, 37–38
media's view of, 19
orders of, 25
origin and ancestors of, 65–66
pace of discovery of, 15–18
polar adaptations of, 212
proliferation of, 103–4
replication of DNA of, 223
seasonal breeding by, 46–47
sexual maturation of, 271
smartest, 17, 29–36
teeth of, *see* teeth, dinosaur
variety in, 20
where to see and dig, 317–20
Dinosaurs From China (Dong and
Milner), 138
Dinosaur Society, 309
Diplodocus, 38, 67, 81, 140, 141, 142,
172, 176, 187, 194
Seismosaurus compared with, 183
Discover, 153
Djadokhta Formation, 248, 249–50
DNA replication, 223
Dockum Formation, 91–92, 93, 94
Dodge Motor Company, 226
*Dodosaurs: The Dinosaurs That
Didn't Make It* (Meyerowitz
and Beard), 23
Dodson, Peter, 39, 47, 49, 50, 99, 141,
161, 188, 189, 194, 272, 274,
275, 278, 279, 305, 309n
Bakker criticized by, 153–56, 162
gender theories of, 40
gigantothermy model of, 163
on Horner, 261
Hunteria criticized by, 154–56
on sauropod size, 190–91
taphonomy and, 41–42
Dong Zhiming, 132–36, 309n
background of, 132–33
Bakker compared with, 133
on catastrophic extinction, 289
at Dashanpu quarry, 138
Dinosaur Project and, 225–28, 240,
242
Mao's friendship with, 134–35
Shunosaurus and, 138, 140
Dromeosaurus, 255

Drumheller conference, 48–49
Dry Mesa site, 169, 172, 176, 182, 184
duckbilled dinosaurs (hadrosaurs),
    221, 257, 261, 304
  crests of, 273–74
  eggs and nests of, 263–64
  environments favored by, 275–76
  jaw motion of, 280
  lifestyle of, 274
  mammals and, 276
  nesting grounds of, 265–67
  sexual dimorphism and, 276
  subgroups of, 273
  teeth of, 273–74, 282
  Weishampel on, 281, 282

Eagles Nest site, 201
earth:
  astronomic cycles and, 114–15
  magnetic field of, 289
  orbit of, 115
Earthwatch, 204
Egg Island site, 264
Egg Mountain site, 263, 264, 270
eggs, dinosaur:
  of bird-hipped dinosaurs, 268
  Dinosaur Project find of, 267
  of duckbills, 263–64
  earliest, 269
  embryos in, 269–72
  from Hell Creek Formation, 267
  of hypsilophodontids, 264
  largest, 268–69
  of lizard-hipped dinosaurs, 268
  of *Maiasaura,* 268
  in Mongolia, 46, 267, 268
  nest arrangement of, 268, 269
  of Sauropods, 268–69
  shape and size of, 268
  structures in, 269
  of *Troödon,* 268
Elliot, David, 145
Elmisaurus rarus, 253
Energy Department, U.S., 180
Erenhot, *see* Iren Dabasu quarry
Eucentrosaurus, 46–47
Europe, 63, 147, 186, 190, 212, 214,
    273, 302
evolution, 16, 19
  of Cretaceous period dinosaurs,
    212, 214

of flowering plants, 153, 214–15
fossil proteins and, 223
Gould on role of chance in, 121
of hip structure, 67–68
"island effects" and, 281
isolation and, 281
jaw motion and, 280
of lizard-hipped dinosaurs, 67–68
"missing link" in, 57, 81–82, 88
"species-packing" and, 283–84
sudden appearance of life and, 54
see also convergent evolution
Ex Terra Foundation, 239
extinction, *see* catastrophic impact
    theory; gradual extinction
    theory; mass extinction theory;
    step-wise extinction theory

family, defined, 24
Farlow, James, 160
  metabolism theory of, 165
Fastovsky, Bruce, 303–4
Ferguson, William Hamilton, 201
ferns, 298
"Fighting Dinosaurs," 232, 249
Fiorillo, Tony, 194
Flaming Cliffs site, 63, 230–31, 232,
    234, 245, 247, 255
  geological formation of, 248–50
Flannery, Tim, 201
flowering plants, 153, 214–15
fluorescence, 181–82
footprints, 113, 116, 118
  herding and, 221
  of iguanodonts, 221
  migration theory and, 221
  speed determined from, 160
foraging rates, 159–60
fossils:
  in Argentina, 59
  biomarkers in, 222–23
  BYU sale of, 197
  chemical tests for, 182
  Chinese policy on, 138
  commercial trade in, 195–97
  distribution pattern of, 221
  DNA replication and, 223
  geographic extent of, 311–13
  "green-stick" fractures in, 47
  herding and distribution of, 221
  lightweight, 222–23

mammal, 112, 230
Martin on examination of, 99–100
measuring age of, 115–16
"missing link" and, 81–82
organic compounds in, 222–23
of plants, 295–96
proteins in, 223
technology and, 180–82
"trace," 247
uranium in, 181, 188–89
France, 146, 267, 268–69
Fryxell Museum, 145

Galton, Peter, 84, 85
gamma-ray spectrometer, 181
Gasosaurus, 125, 186–87
Gasosaurus constructus, 144–45
gastroliths, 189
Gauthier, Jacques, 50, 72–73, 76, 88, 99
  cladistic study of, 69–70, 85
  dinosaur-bird debate and, 85, 86
  sauropod characters named by, 191
Gem & Mineral Show, 195
gender theory, 40
Genghis Khan, 227
genus, defined, 24
geochemical dating, 116
Geological Society of America, 293
Geological Survey, U.S., 116, 218, 293, 304
George, Eldon, 110
"Gertie" (dinosaur), 62
gigantothermy, 163
Gillette, David, 50, 182, 185
  background of, 179
  fossil collecting opposed by, 195, 197
  gastroliths discovered by, 189
  Seismosaurus and, 179–80, 183–84
Gilmore, Charles, 264–65, 267
giraffes, 141
Gobi Desert, 16, 38, 49
  Andrews's expedition to, 226, 228–31
  "Fighting Dinosaurs" find in, 232, 249
  geography of, 226–27
  loess of, 227
  mammal fossils in, 230
  Polish expedition to, 232–34
  Soviet research in, 235

Gondwana, 24
  splitting of, 186
Gorman, James, 162
Gould, Stephen Jay, 68, 82, 150, 229, 291
  Bakker assessed by, 152
  on catastrophic impact theory, 288
  on Cuvier's work, 37
  on paleontologists, 288
  on quality of scientists, 18
  on role of chance, 121
graben, tectonic, 248–49
gradual extinction theory, 218–20, 297
  see also catastrophic impact theory;
    mass extinction theory;
    step-wise extinction theory
Granger, Walter, 228
Great Basin, 188
Great Basin Naturalist, The, 174
Great Britain, 139, 146, 213, 214
Great Dinosaur Rush, 37–38
greenhouse effect, 289, 296
"green-stick" fractures, 47
Gregory, R. T., 210
Griffin, Emily, 154
Grigorescu, Dan, 281–82
growth of dinosaurs:
  phases of, 163–65
  rates of, 157–58, 163–64, 271
growth rings, 157, 165, 220

Hadrosaurines, 273–74
hadrosaurs, *see* duckbilled dinosaurs
Hagan, Roland, 180
Hager, Mick, 261
Hallett, Mark, 194
Harvard University, 150, 152, 169
Hateg Basin, 282
Hecht, Max, 85
Hedin, Sven, 231, 242
Heilmann, Gerhard, 82–83
Hell Creek Formation, 50, 218, 304
  dinosaur diversity in, 302–3
  first eggshells from, 267
Hennig, Willi, 68
herding, 151, 188
  Currie's study of, 46–47
  footprints and, 221
  fossil distribution patterns and, 221
Herrera, Victorino, 62

Herrerasaurus ischiqualastensis, 62,
  66, 103
  discovery of, 70–73
  jaw of, 73
  lizard-hipped dinosaurs and, 72
  "oddball characteristics" of, 72
Hesperornis, 84
hibernation, 200, 212
Hickey, Leo, 297–98
"high-energy aggression," 162
Hildebrand, Alan, 291–92
Hirsch, Karl, 18, 268
Hiyan Tong, 146
holotype, 176
homeothermy, mass, 162, 165
Hoover, Herbert, 231
horned dinosaurs (ceratopsians), 230,
  265, 276, 304
  Asia's lack of, 253–54
  beak of, 277
  behavior of, 278–79
  families of, 277
  habitats preferred by, 279
  horn and hood variety, 277–78
  Serono's reclassification of, 276–77
  sex distinctions in, 278
Horner, John R. "Jack," 18, 50, 110,
  162, 257, 268, 271, 272, 274,
  275, 279, 280, 309*n*
  background of, 259–61
  commercial collectors and sites of,
  196
  Cretaceous habitat as imagined by,
  258
  Dodson on, 261
  embryos found by, 270
  first eggs and nests found by, 261,
  263–64
  "genius" grant awarded to, 261
  growth patterns studied by, 165
  Landslide Butte site located by,
  264–65
  Maiasaura named by, 264
  on mass homeothermy, 165
  MSU's hiring of, 261
  nesting grounds found by, 265–67
  nests discovered by, 45–46
  Peterson and, 163–65
  species formation theory of, 284–85
  Troödon discovered by, 29–36
  Weishampel and, 282–83

"Hot & Cold Running Dinosaurs"
  (Bakker), 151
Hotton, Nicholas, III, 92, 96, 98,
  141–42, 309*n*
  on Chatterjee, 90, 99
  dinosaur migration as favored by,
  220–21
Huayangosaurus, 142–44, 147
Huene, Friedrich von, 59, 62
  dinosaur migration concept of,
  220
Hunteria, 154–55
Hutchinson, Howard, 218, 221
Hypsilophodon, 196
Hypsilophodonts, 187, 205, 258, 282
  eggs and nests of, 264
  embryos of, 270–71
  flowering plants and, 215
  kangaroo-like, 206–7
  polar habitat adaptations of, 210,
  212
  at Proctor Lake site, 213
  teeth of, 213, 271

Ichthyornis, 84
  teeth of, 86
ichthyosaurs, 53
Iguanodon, 37, 38
  discovered in Belgium, 213–14
Iguanodon bernissartensis, 213–14
Iguanodon mantelli, 214
iguanodonts, 205, 273
  disappearance of, 215
  footprints of, 221
Illustrated Encyclopedia of Dinosaurs,
  The (Norman), 319*n*
impact extinction, *see* catastrophic
  impact theory
"impact winter," 296
India, 16, 38, 46, 89, 92–93, 214, 267,
  289, 302
Indian Statistical Institute, 89
Indricotherium, 230
Institute of Vertebrate Paleontology
  and Paleoanthropology (IVPP),
  131, 132, 138
  achievement of, 136
  cultural revolution and, 134–36
  Lucas on, 134–35
  Sereno on, 135
intelligence, brain size and, 210

Iren Dabasu quarry, 226, 232
    Andrews's visit to, 228, 229–30
    Dinosaur Project at, 251–55
iridium, 44, 116, 292, 295, 305
    catastrophic impact theory and,
        119, 290, 298, 299
    in comets, 294
    at K-T Boundary, 119, 298, 299
    Manicouagan impact site and, 119
Ischigualasto Formation, 51, 57–62,
        76, 136, 179
    described, 58–59
    Herrerasaurus discovered in, 70–73
"island effects," 281

'Jack's Birthday Site," 50
Jacobs, Louis, 213
jaw motion, 280
Jefferson, Thomas, 37
Jensen, Jim, 59, 169–76, 178, 179, 182,
        184, 197, 267
    background of, 169
    at Dry Mesa, 172
    Hypsilophodon lost by, 196
    McIntosh acknowledged by, 192
    Madsen's criticism of, 185
    professional shortcomings of,
        174–76
    storage of finds of, 185–86
    Ultrasaurus as described by, 174–76
    zoological nomenclature and,
        172–74
Jensen, Ron, 172
Jerzykiewicz, Tom, 248–50
Jiangjunmiaosaurus, 145
Johnson, Kirk, 297
Johnson, Rolf, 159
Johnston, Paul, 247
Jones, Eddie, 170
Judith River formation, 40, 160, 276,
        279, 304
Jurassic, George, 181
Jurassic Park (Crichton), 222
Jurassic period, 81, 117, 118, 121,
        140, 146, 147, 186–89, 190, 213,
        214
    American West in, 187–88
    ascendant dinosaur forms of,
        186–87
    climate of, 186
    defined, 24

environment of, 187–88
    Morrison Formation created in, 187
    Pangaea's breakup in, 186
    proliferation of giants in, 103–4
Jurassic-Cretaceous Boundary, 137,
        153

Kepler, Johannes, 114
Kielan-Jaworowska, Zofia, 232, 252
    Andrews refuted by, 234
    Asian-American migration doubted
        by, 235
Kim, Haang Mook, 174
Kitching, James, 89, 269
Koonwarra site, 210
Kurzanov, Sergei, 253, 309*n*

Laboratory of Vertebrate
        Paleontology, 131
labyrinthodont, 206
Lagosuchus, 73–76
L'Aigle, France, 119
lambeosaurines, 273–74, 276
Landslide Butte site, 265, 267
"last day phenomenon," 252
Laurasia, 24
    splitting of, 186
Lawrence Berkeley Laboratory, 290,
        294
Leaellynasaura, 207
    large brain of, 210
Lehman, Thomas, 278
Leidy, Joseph, 37
Leitch, Andrew, 272
Leptoceratops, 285
Life, 151
Li Ming, 136
"living fossils," 212
lizard-hipped dinosaurs
        (saurischians):
    cladistics and, 69
    defined, 25
    eggs of, 268
    evolution of, 67–68
    Herrerasaurus and, 72
    proliferation of, 103–4
Lockley, Martin, 221
loess, 227
Long, John, 201
Los Alamos National Laboratory, 180,
        181, 182, 223

Lucas, Spencer, 48–49
  Dong as viewed by, 133–34
  on IVPP, 134–35
"lucky break" theory, 105–6
Lunar Planetary Laboratory, 291–92
lungfish, 206

Maastrichtian period, 296, 301
MacArthur, Robert, 281
MacArthur Foundation, 261
McIntosh, John "Jack," 18, 49, 186,
      188, 189
  Apatosaurus identified by, 193
  background and personality of, 192
  Bakker's recollection of, 192
  "Brontosaurus" reexamined by, 193
  described, 191–92
  Jensen's acknowledgement of, 192
  on sauropod evolution, 140
  sauropod interest of, 192–93
  Shunosaurus viewed by, 139–40
  on Supersaurus identification,
      184–85
McKenna, Malcolm, 255
Madsen, Jim, 185, 197
magnetometer, 181
Maiasaura, 45–46, 163, 265, 267, 276,
      284
  eggs of, 268
  embryos of, 270–71
  growth phases of, 163–64
  Horner's naming of, 264
Makela, Robert, 45, 263, 270
Mamenchisaurus, 242
Mamenchisaurus constructus, 132,
      141
mammals, 65, 302
  duckbill dinosaurs and, 276
  Gobi fossils of, 230
  jaw-joint of, 66
  largest, 230
  Richmond fossil of, 112
Mandschurosaurus, 131
Manicouagan impact site:
  absence of iridium at, 119
  mass extinction and, 118–21
  shattered quartz at, 119
Manley, Kim, 182
  Seismosaurus's world as compiled
      by, 188–89
Manson crater, 294

Mantell, Gideon, 37, 195
Mantell, Mary Ann, 37
Mao Tse-tung, 131, 134
Marsh, Othniel Charles, 37–38, 39, 136
  "Brontosaurus" and, 193
marsupials, 301
Martin, Larry:
  dinosaur-bird debate and, 85–86
  on examination of fossils, 99–100
  Protoavis endorsed by, 96–97
Martin Marietta Energy Systems, 180
Maryanska, Teresa, 235
mass extinction theory, 44
  acid rain and, 295
  arguments against, 300–303
  asteroids and, 293
  climate and, 115, 289
  epidemics and, 153
  "lucky break" theory of, 105–6
  Manicouagan site and, 118–21
  periodicity of, 294
  in Permian era, 66
  plant fossil evidence for, 295
  of plants, 298
  small animals and, 118
  survivors of, 301–2
  at Triassic-Jurassic boundary, 104–6,
      107, 112, 117
  volcanism and, 299–300
  see also catastrophic impact theory;
      gradual extinction theory;
      step-wise extinction theory
mass homeothermy, 162, 165
Maurrasse, Florentine, 292
Mayr, Ernst, 281
Megalosaurus, 37, 144–45
Mesozic era, 59, 66, 67
  periods of, 23–24
metabolism, 43, 147, 151, 157
  body size and, 162–63
  Farlow's theory of, 165
  gigantothermy model of, 163
  growth rates and, 271
  polar conditions and, 220
  see also warm-bloodedness
Meyerowitz, Rick, 23
meteorite detection, 118–19
Michel, Helen, 290
Microhadrosaurus, 134
migration, 200, 289
  Alaskan evidence of, 219

Currie's support of, 221
footprints and, 221
Hotton's arguments for, 220–21
Huene's concept of, 220
warm-bloodedness and, 220–21
see also Asian-American migration
Milankovitch cycles, 114–15
Miles, Cliff, 176–77
Milk River site, 271
Miller, Wade, 197
Milner, Angela, 138
Milwaukee Public Museum, 50, 159,
    303
"missing link," 57, 81–82, 88
"Missing Link Expedition," 229
Molnar, Ralph, 201, 309*n*
Monash University, 201, 203
Mongolia, 16, 20, 63, 225, 226, 230,
    231, 240, 270, 302
    dinosaur eggs in, 46, 267, 268
    Troödon found in, 36
    see also Gobi Desert
monsoons, 115
Montana, 16, 20, 28, 42, 46, 163, 187,
    214, 221, 257–58, 261, 267–68,
    272, 277, 298
    Cretaceous floods in, 279, 283–84
Montana State University, 164, 165
    Horner hired by, 261
Montanoceratops, 285
Morgan, J. P., 231
Morrison Formation, 181
    forming of, 187
    Serengeti compared with, 188
    uranium in, 188–89
mosasaurs, 301
Muller, Richard, 294
multituberculates, 301
Museum of Natural History (Beijing),
    135
Museum of the Rockies, 18, 196, 257,
    261, 264
Mussaurus, 269
Muttaburrasaurus, 213

National Academy of Science, 196
National Geographic, 120, 299
National Geographic Society, 96, 98,
    111
National Science Foundation, 50, 70,
    90, 184, 185

Nature, 153
Nature Conservancy, 264
Nemesis (dark star), 294
nesting ground, 265–67
Newark Supergroup, 110, 112–17, 118,
    120
    age of fossils in, 115–16
    ancient lakes of, 114–16
    geology of, 113–14
    volcanic activity and, 116
New Mexico, 187, 221, 250, 298
New Mexico Museum of Natural
    History, 48, 179, 184
Newsweek, 29, 49, 165
Newton, Isaac, 114
New York Times, 98, 164, 169
Noble, Brian, 238–39, 240, 245, 247
nodosaurs, 187, 214
nodule zone, 249–50
nomenclature, zoological, 172–74
Nopsca, Franz Baron von, 164,
    280–82
Norman, Dave, 280, 282, 309*n*
North America, 16, 17, 18, 120, 140,
    143, 147, 186, 190, 218, 246,
    251, 253–54, 273, 277
    Cretaceous floods in, 283–84
    Gobi finds and, 234–35
    Newark Supergroup of, 110, 112–17
North Slope, Alaska, 215–16, 218
Nova Scotia, 20, 106, 113, 117, 186
    ancient climate of, 118
Nuclear Winter, 16, 44, 120, 201, 294,
    296

obliquity cycle, 115
Ocean Point site, 216
Oklahoma Museum of Natural
    History, 185
Olsen, George, 230
Olsen, Paul, 106–21, 304
    astronomical cycles and, 114–15
    background of, 113
    Benton's theory rejected by, 117
    catastrophic impact theory favored
        by, 117–18, 120–21
    Padian and, 113
    Triassic-Jurassic boundary dated by,
        116–17
    Weishampel on, 112
Omeisaurus, 131, 135

Oort Cloud, 294
Origin of Birds, The (Heilmann), 82
Origin of Species (Darwin), 81
ornithischians, *see* bird-hipped
    dinosaurs
Orodromeus, 268
Orodromeus makelai, 270
Osborn, Henry Fairfield, 89, 254
    Asia origins theory of, 229
Osmolska, Halszka, 235–36, 252–53,
    309*n*
osteocalcin, 223
osteons, 164
ostrich-mimic dinosaur, 100–101, 235
Ostrom, John, 42–43, 49, 161
    Bakker and, 152
    Deinonychus described by, 157
    Deinonychus found by, 214
    dinosaur-bird theory and, 83–84,
        152
    dinosaur postures tested by, 159
    Protoavis and, 96, 98
Otway National Park, 203
Oviraptor, 230, 255
Owen, Richard, 37
oxygen-isotope ratios, 210

pachycephalosaurs, 252, 304
Pachyrhinosaurus, 218, 221
Padian, Kevin, 57, 84–85, 88, 106, 120,
    288–89
    Chatterjee criticized by, 99
    Olsen and, 113
Paleobiological Institute of the Polish
    Academy of Sciences, 232
paleomagnetism, 146
Paleotherium, 37
Pangaea, 23, 81
    breakup of, 186
    climate of, 53–54
    monsoons and, 115
Parrish, Judith, 298
Parrsboro, Nova Scotia, 106–7
partitioning of resources axiom, 221
Peabody Museum of Natural History,
    18, 113, 192
Penfield, Glen, 292
Perle, Artangerel, 267
Permian period, 54
    climate of, 66
    great extinction in, 66

Peterson, Jill, 163–65, 307
Petrified Forest, 62, 241
*Philosophical Transactions of the
    Royal Society of London, The,*
    98–99
Physical Review, 192
pinacosaurs, 247
Pinacosaurus, 251
plesiosaurs, 53, 206, 301
Plot, Robert, 36
Poinar, George O., 223
polar habitats, 210, 212
Polish-Mongolian Dinosaur Dig, 248
polymerase chain reaction, 223
Popigay crater, 291
Post, C. W., 91
Postosuchus, 93
predators, 54, 104, 144
    prey ratios to, 160
Princeton University, 164, 259, 261, 282
Prinn, Ronald, 294–95
Proctor Lake site, 213
prosauropods, 118, 269
Protoavis, 87–89, 93–100, 103
    Archaeopteryx and, 87, 93, 95
    Bakker on, 97
    Chatterjee's analysis of, 94–97
    Chatterjee's discovery of, 87–89,
        93–94
    Chatterjee's treatise on, 98–99
    described, 94
    dinosaur-bird debate and, 86–89,
        93–100
    Gauthier's rejection of, 97–98
    Martin's endorsement of, 96–97
    Ostrom and, 96, 98
    Troödon and, 95, 97
Protoceratops andrewsi, 230, 232, 234,
    235, 245, 247, 251, 254, 255,
    269–70, 277
psittacosaurs, 254, 277
pterosaurs, 83, 206, 241, 301
    ankle bones of, 66–67
Puchezh-Katunki crater, 121

Qongqing Museum, 132, 135
Quest for Fire (film), 40
quicksand, 249

radar, ground-penetrating, 181
Reig, Osvaldo, 59–62

Repenning, Rip, 218
rhynchosaurs, 54
Rich, Leaellyn, *Leaellynasaura* named
    for, 207
Rich, Tom, 201–6, 210, 212
Richmond, Va., 112
Rich-Vickers, Pat, 201, 207, 210, 212
Ricqlès, Armand de, 157–58, 164
Rife, Jill, 165
Rigby, Keith, Jr., 302
Romania, 281
Romer, Alfred, 59, 73, 169
Romer Prize, 164
Royal Ontario Museum, 111, 116
Royal Tyrrell Museum of
    Paleontology, 29–32, 136, 221,
    238, 245
    SVP conventions at, 47–51
Russell, Dale, 39, 141, 220, 304, 308
    Dinosaur Project and, 240, 242

Sagan, Carl, 291
St. Mary's Formation, 283, 284–85
Sampson, Scott, 278–79
Sandia Laboratories, 181
saurischians, *see* lizard-hipped
    dinosaurs
sauropods, 131, 153
    archaic form of, 191
    architectural peculiarities of, 191
    characteristics of, 191
    defined, 25
    Dodson on size of, 190–91
    dorsal plates of, 142–44
    dorsal spines and, 139
    eggs of, 268–69
    feeding strategy of, 160–61
    feet of, 190–91
    gastroliths in, 189
    geographic range of, 190
    head elevation of, 141–42
    lifespan of, 271
    long-necked, 141
    McIntosh and, 192–93
    McIntosh on evolutionary role of,
        140
    success of, 190
"Scholar of the House" project, 151
Schwartz, Heidi, 182
Schwarzenegger, Arnold, 153
Scientific American, 152

sedimentology, 248–51
Seismosaurus, 50, 179, 181, 182, 186,
        194
    Diplodocus compared with, 183
    discovery of, 179–80
    environment of, 188–89
    gastroliths and, 189
    naming of, 183
    "Selling Fossils," 179, 195–97
Sereno, Paul, 50–51, 54, 137, 142, 238,
        254, 309n
    background and career of, 62–64
    ceratopsian characters identified by,
        276–77
    cladistics study by, 69, 235
    dinosaur-bird debate and, 84
    fossil-finding knack of, 65
    Herrerasaurus discovered by,
        71–72
    Ischigualasto Formation visit of,
        57–59, 70–73, 76
    Lagosaurus sought by, 76
    on working at IVPP, 135
Sevens, Paul, 93
sexual dimorphism, 143, 214
    in ceratopsians, 278
    in duckbills, 276
Shantungosaurus giganteus, 134
shatter cones, 118
Sheehan, Peter, 303–4
shocked quartz, 93, 292
    at Manicouagan site, 119
    at Wassons Bluff, 120
Shoemaker, Caroline, 293
Shoemaker, Gene, 293
Shubin, Neil, 111, 119
Shunosaurus, 124, 125, 128
    described, 138–39
    Dong and Xiao and, 138, 140
    fossil skulls of, 139–40
    McIntosh's viewing of, 139–40
    tail of, 140
Siebenbürgen fauna, 281
Sinkiang, China, 240
    dinosaur finds in, 242
    fossil grove found in, 241
Slippery Rock site, 204
Society of Vertebrate Paleontology
        (SVP), 48–51, 100, 164, 197,
        270, 276
sonic images, 180–81

South Africa, 46, 214, 267, 269
South America, 16, 17, 118, 190, 206, 273, 277, 302
  geological isolation of, 59
  sauropod evolution in, 214
Spall, W. Dale, 223
species and speciation:
  defined, 24, 67
  habitat and formation of new, 258
  Horner's theory on formation of, 284–85
  "island effects" and, 281
  rate of, 162
  Weishampel's theory on formation of, 284–85
"species-packing," 283–84
spectrometer, gamma-ray, 181
speed of dinosaurs, 151, 159
  calculated from footprint, 160
Spicer, Robert, 222, 298
Spotila, James, 163
Stadtman, Ken, 172, 176
Stahmer, Bob, 149
Stanley, Steve, 299
State Museum (Ulan Bator), 232, 235
Staurikosaurus, 62
stegosaurs, 25, 153, 304
Stegosaurus, 19, 38, 143, 187
step-wise extinction theory, 299–300
  see also catastrophic impact theory; gradual extinction theory; mass extinction theory
Sternberg, Charles M., 263, 305
Sternberg family, 39
styracosaurs, 284
Sues, Hans-Dieter, 111, 112
Supersaurus, 81, 172, 174, 176, 178, 183, 194
  McIntosh on identification of, 184–85
SVP, see Society of Vertebrate Paleontology
Swinburn, Nicola, 293
synapomorphies, 68
Szechwan, China, 28, 124, 125, 132, 146
  geology of, 136–38

Tang Zhilu, 34, 240
taphonomy, 40–41, 46
  Dodson and, 41–42

tarbosaurs, 235
Tarsitano, Sam, 85
technology, digging and, 180–82
Technosaurus, 92, 93
tectonic graben, 248–49
teeth, dinosaur:
  of duckbills, 273–74, 282
  of embryos, 271–72
  of hypsilophodonts, 213, 271
  of Ichthyornis, 86
tektites, 292
temperature regulation, 162–63
Tertiary period, 289, 296, 302
Texas, 65, 187, 250, 277
  Proctor Lake site in, 213
"Texas Fossil May Be Birds' Oldest Ancestor" (Wilford), 98
Texas Tech University, 90, 92
thecodonts, 83
therapsids, 66
Therizinosaurus, 232, 234
theropods, 81, 206, 235, 242, 304
  defined, 25
"throw away leaf," 297
Torosaurus, 159
"trace fossils," 247
trackways, 110
  migration theory and, 221
Transylvania, 16, 46, 267, 281–82
Triassic-Jurassic boundary, 186
  age of, 116–17
  catastrophic impact theory and, 119
  lake formation at, 116
  mass extinction at, 104–6, 107, 112, 117
Triassic period, 66, 70, 81, 91, 96, 103, 117, 118, 120, 121, 134, 137, 186, 269
  defined, 23–24
  Ischigualasto Formation and, 58–59
Triceratops, 46, 159, 230, 254, 265, 277, 304
tritylodont, 241
Troödon formosus, 206, 218, 240, 264, 271
  Currie and, 29–36
  dinosaur-bird debate and, 36, 84
  eggs of, 268
  embryo of, 270
  Horner's discovery of, 29–36
  Protoavis and, 95, 97

Tsingtaosaurus, 132
Tunke, Darren, 278
tunneling burrows, 247
Twain, Mark, 19
Two Medicine River site, 264, 283–84, 285
Tyrannosaurus rex, 18, 19, 25, 39, 67, 87, 93, 232, 234, 235, 304
 speed of, 160
Tyrrell, Joseph Burr, 39

Ulan Tsonchi site, 231, 245, 251
Ultrasaurus, 81, 178, 184, 185, 187, 193, 194
 Jensen's description of, 174–76
Ultrasaurus macintoshi, 192
Ultrasaurus tabriensis, 174
ultraviolet light, 181–82
uranium, 181
 in Morrison Formation, 188–89
 sale of fossils and, 196
Utah, 187, 188, 196, 197

Velociraptor, 232, 242, 247, 255
Versey, Brian, 176–77
Victoria Museum, 201, 203, 204
volcanism, 289, 298
 Basu's theory of, 300
 mass extinction and, 299–300
 Newark Supergroup and, 116

Walker, Alick, 83, 86
Walker, William, 214
Walter Kidde Dinosaur Park, 113
Ward, David, 63–64
warm-bloodedness, 42–43, 83, 156–65, 200, 288, 302
 adaptation for speed and, 159–60
 Bakker's advocacy of, 149, 151, 153, 156–57
 behavior and, 161–62
 dinosaur embryos and, 270–71
 feeding strategy and, 160–61
 growth rates and, 157–58, 163–64
 high blood pressure and, 142
 migration and, 220–21
 polar habitats and, 212
 predator-prey ratios and, 160
 Ricqlès's doubts on, 157–58
 speciation rate and, 162

vascular system and, 162
 see also metabolism
Wassons Bluff, 112
 age of, 116–17, 120
 catastrophic impact and, 118
 climate of, 118
 described, 107–11
 fauna of, 117
 pollen findings at, 118
 shocked quartz at, 120
Weeks, Kent, 181
Weishampel, David, 112, 191, 258, 270, 271, 275, 279, 309*n*
 background of, 279–80
 on duckbills, 281, 282
 Horner and, 282–83
 jaw motion studied by, 280
 on Olsen, 112
 species formation theory of, 284–85
 Transylvanian fossils studied by, 281–82
Wesleyan University, 18, 192
Wilford, John Noble, 98, 169
Williams, Thad, 213
Willow Creek anticline, 265
Wilson, E. O., 281
Wolfe, Jack, 295–97, 298
Winkler, Dale, 213
Witten, Alan, 180, 181

Xian University, 136
Xiaosaurus, 187
Xiaosaurus dashanpensis, 142
Xiao Xi Wu, 133, 135, 136, 142, 144, 145–46, 147, 149
 at Dashanpu quarry, 138
 Shunasaurus and, 138, 140

Yale University, 84, 113, 152, 193, 261
Yangchuanosaurus, 134
Yang Zhong-jian (C. C. Young), 131–32
Yenching Library, 150
Yu Chao, 136

Zdansky, Otto, 131
Zhao Xijin, 242
Zheng Jiajian, 242
Zigong Dinosaur Museum, 125, 128
zoological nomenclature, 172–74

THE TIME OF

Dinosaurs ruled the earth for 150 million years through three different periods in earth's history. The land, the climate, and the dinosaurs changed greatly in that time.

TRIASSIC

245 million to 208 million years ago. All the land was joined in one continent called Pangaea. The climate was warm and often rainy. The first dinosaurs spread across the land in this time. They were hunters, less than 10 feet long. Pterosaurs, flying reptiles, first took to the air.

Earth age 4.6 billion years ago

First single cell life 3.5 billion years ago

TIME LINE